D1632212

Diesels Afloat

Diesels Afloat

PAT MANLEY

John Wiley & Sons, Ltd

Published under the Fernhurst imprint by John Wiley & Sons Ltd, The Atrium, Southern Gate, Chichester, West Sussex PO19 8SQ, England

Telephone (+44) 1243 779777

Email (for orders and customer service enquiries): cs-books@wiley.co.uk
Visit our Home Page on www.wiley.com

Other Wiley Editorial Offices

John Wiley & Sons Inc., 111 River Street, Hoboken, NJ 07030, USA

Jossey-Bass, 989 Market Street, San Francisco, CA 94103-1741, USA

Wiley-VCH Verlag GmbH, Boschstr. 12, D-69469 Weinheim, Germany

John Wiley & Sons Australia Ltd, 42 McDougall Street, Milton, Queensland 4064, Australia

John Wiley & Sons (Asia) Pte Ltd, 2 Clementi Loop #02-01, Jin Xing Distripark, Singapore 129809

John Wiley & Sons Canada Ltd, 6045 Freemont Blvd, Mississauga, ONT, L5R 4J3

Wiley also publishes its books in a variety of electronic formats. Some content that appears in print may not be available in electronic books.

Index compiled by Alan Thatcher

Anniversary Logo Design: Richard J. Pacifico

Library of Congress Cataloguing in Publication Data

Manley, Pat.
Diesels afloat / Pat Manley.
p. cm.
Includes bibliographical references and index.
ISBN: 978-0-470-06176-3 (pbk. : alk. paper)
1. Marine diesel motors. I. Title.
VM770.M347 2007
623.87´236–dc22 2006038742

British Library Cataloguing in Publication Data

A catalogue record for this book is available from the British Library

ISBN 978-0-470-06176-3 (PB)

Typeset in 9/12 pt Swiss 721 by Thomson Digital
Printed in Italy by Printer Trento, Trento
This book is printed on acid-free paper responsibly manufactured from sustainable forestry in which at least two trees are planted for each one used for paper production.

Contents

Preface

In 2005, The UK's RNLI lifeboats were launched 8421 times for all causes. Of these, 1710 launches were to vessels suffering machinery failure, 1298 being to leisure craft (387 of these were sailing vessels and 911 were power craft). These numbers don't include failures dealt with by professional breakdown services or friendly tows.

Some knowledge of the operation of diesel engines and their systems, their maintenance and the ability to carry out simple repairs would have reduced the need for the rescue services to attend these craft significantly.

The RYA's desire to improve this situation led to the introduction of the RYA Diesel Engine Course in 1996, and I became one of its earliest instructors. *Diesels Afloat* incorporates much of the feedback from yachtsmen who have attended my RYA Diesel Engine courses since their inception, and covers the complete syllabus in depth. It answers the questions that I'm asked most frequently and which aren't answered in other books. However, there's much additional information that I consider essential for a proper understanding of a diesel engine's working and its operation by the skipper.

Troubleshooting is often covered by tables giving no definitive answer. I believe that a proper understanding of the diesel engine makes these tables superfluous and instead use the much more helpful descriptive text combined with a proper understanding of the various systems.

A boat's engine cannot be considered alone. There's a need to convert its power to propel the boat and this part of the system is often considered a black art. As I am often being asked about propellers, both by students and *Practical Boat Owner* magazine readers, I have devoted a chapter to this subject.

Pat Manley

Introduction

During the 1930s, with the increasing popularity of sailing cruising yachts as a leisure pursuit, fitting an auxiliary engine became common. The unwillingness of these small petrol engines to start, and their general unreliability, meant that their use tended to be confined to occasions when there was no wind, and even then, many yachtsmen sailed as if they had no engine. They were, in all senses of the word, auxiliary engines and were hated and distrusted by many yachtsmen.

As auxiliary engines became more common, a rule of thumb developed such that their size for any given yacht was about 2 horsepower for every ton of displacement. This was adequate for getting home when the wind was too light to sail, but was not too much use for battling into wind and sea. But as it was a sailing boat after all, that did not matter.

When small, reliable marine diesel engines became available in the 1960s, coupled with the need to get home to go to work, engine powers were increased, such that 4 horsepower per ton of displacement became the norm. This allowed sailing boats to be motored into wind and wave and make progress, and in calm water they could achieve their maximum 'hull' speed. That is, they could go as fast as was economically sensible under power. Displacement motorboats could use the same rule of thumb, but as they had no alternative means of propulsion, were often given a bit more power.

Into the 1990s, there came a tendency by some boat builders to 'up' the power to around 6 horsepower per ton displacement for sailing boats. In many ways this was because although the sailing performance of most current sailing yachts is very good, they are often used as motor sailing boats. Many owners reach for the engine starter if the boat speed drops below 5 knots or if they have to go to windward. However, this relative overpowering brings a hidden cost. Diesel engines must be worked hard, as we shall demonstrate later, so the boat must be cruised at a speed higher than its 'hull speed', with its attendant large increase in fuel consumption.

Some motor cruisers have a similar problem to some sailing boats, in that they are overpowered for the conditions in which they will be used. Displacement motorboats would be adequately powered at 4 to 6 hp per ton displacement, which is generally fine. Planing boats need much more and, generally speaking, are 'cruised' at high power and high speed. It is these boats that experience problems when run at low speed, as may be dictated by inland waterway speed restrictions. The need for turbo-charged diesels to be run under load is even more demanding than for those normally aspirated. Boats that are going to be used only on inland waterways should be powered accordingly. This is sometimes seen as a potential liability when it comes time to sell the boat, so is often ignored. Twin engine boats can be run safely on one engine at a time, so that the engine can be run at higher power, swapping engines to equalise the hours run. Be aware though, that to save production costs, some twin engine boats have a power steering pump fitted to just one engine, so single engine running on the 'wrong' engine can produce interesting results. I was assured recently by a salesman that a twin engine boat had a pump on each engine, despite it being obvious that only one was fitted. Running the engines one at a time showed the salesman that he was in error.

Rather than the sailing boat's engine being an auxiliary, it has become the alternative means of propulsion. Obviously for a single engine motorboat, it's the only means of propulsion, whereas a twin engine motor boat still has an engine should the other fail, provided that you can still steer the boat.

Because car engines have become so reliable, many boat owners take the same attitude towards their boat's engine as they do their car's. But there's a significant difference – the boat is at sea, where there can be significant corrosion problems, and you can't just pull to the side of the road and stop as you can on the road. Additionally, because the boat engine gains little annual mileage and often isn't even warmed up before stopping, it can suffer from unsuspected neglect.

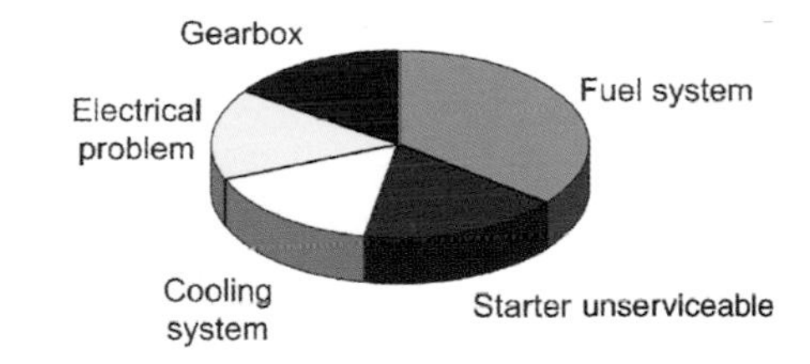

Top five reasons behind machinery failure for 2005

Courtesy of the RNLI.

RNLI statistics for 2005 reveal the various causes of failure.

This book will look at all the aspects of the various parts of the engine and its systems, maintenance, troubleshooting and use of your marine diesel engine so that you can get the best out of it with the least chance of its letting you down.

The engine's handbook is there to be read. It's quite amazing the interesting and useful things you will find in it. There's a whole generation of Volvo Penta 2000 series engine owners who just do not know how the engine should be started from cold, despite clear instructions being included in the handbook. As I tell all my students – READ THE HANDBOOK!

Pat Manley

The Diesel Engine

HOW IT ALL STARTED

Rudolf Diesel was granted the first patent for a diesel engine in 1892, when petrol engines were in their infancy. Whereas petrol engines could be built small enough to be put in a motor car, the diesel engine was on a different scale completely. Early examples were 3 metres tall!

Although diesels were used in German flying boats and even Zeppelins in the 1930s, they were really too big and heavy and were not considered a success.

It was not until the late 1950s that any real success was achieved in building small, relatively lightweight diesels for use in small leisure craft. These were one-, two- and three-cylinder engines revving at around 2300 rpm and developing from 7 to 35 hp. They were quite heavy and bulky, but the smallest could be fitted into a 20-foot boat.

In 1970 Petter produced a 6 hp single-cylinder engine built mainly from aluminium and derived from one of their small industrial units. This was very compact, light in weight and revved at 1500 rpm. This was quickly followed by a two-cylinder 12 hp version.

Larger boats needed more power, and by this time there were a number of smaller diesel-powered cars on the market, some of which were marinised by independent companies to give engines producing 35 to 45 hp. Larger motor and work boats used marinised truck engines.

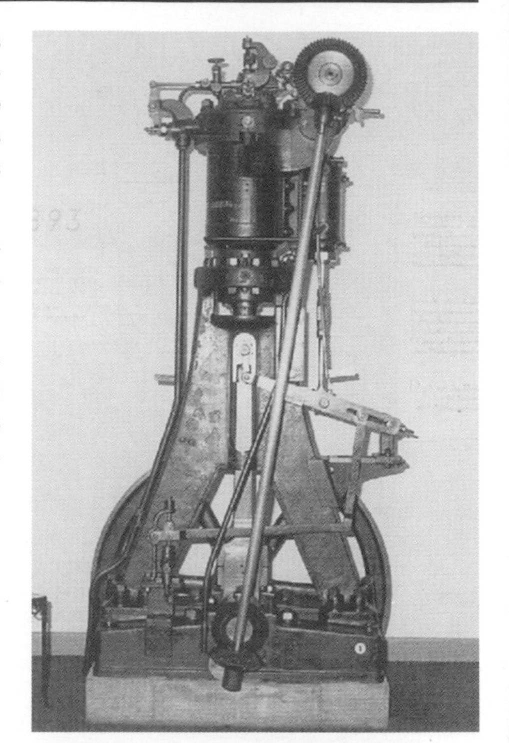

Planing motor cruisers need lots of power and relatively light weight, and it's here that the modern turbocharged truck engine plays its part.

All modern marine diesels destined for the leisure market are marinised versions of automotive or industrial engines. This marinisation may be carried out by the original engine builder or by an independent marinising company.

THE MODERN DIESEL ENGINE

Compared with its forbears, the modern diesel is light in weight and relatively high revving. It is in all ways comparable to the modern petrol engine but more economical to run and a little more expensive to buy.

Modern diesels range from around 10 horsepower right upto 1000 or so in the leisure engine ranges. Low-powered engines up to around 40 hp are simple in operation, with no electronic 'goodies'. Above 100 horsepower, leisure engines rely much more on electronic engine control and methods of augmenting their power. Between 40 and 100 hp it's an either/or situation.

The type of engine we use should be dictated by the use to which it will be put.

Displacement hulls

Where the boat's speed is limited by its waterline length, relatively low power is required. The old rule of thumb was 2 hp per tonne displacement. Much more realistic in these days of needing to get home to go to work would be 4 hp per tonne. Many builders seem to be offering as much as 6 hp per tonne or even more, but this brings with it problems of high fuel consumption and engines that are run at far too low a power for normal cruising. Diesels need to be worked hard, so it's no good saying that I won't use all that extra power that I've installed 'just in case'. If you don't work them hard you are storing up problems for later, and sometimes sooner, in their life.

Displacement motor boats are normally cruised at a constant speed and the engine is in use all the time. The engine gets warmed up properly and as long as it doesn't have too many hp per tonne, it gets a reasonably easy life.

A sailing boat's engine has a much harder time, as often it doesn't reach normal running temperature before it's stopped. It's also often used at relatively low power when 'motor sailing'. These conditions are not good for a diesel, and even less good if it's turbo-charged. Unless there's just no suitable non-turbo engine of the power required, I'd suggest avoiding a turbo-charged engine in a sailing yacht.

A modern 36-foot yacht weighing 6 tonnes needs 24 hp by the 4 hp per tonne rule. This would give a cruising speed of 6.6 knots with a fuel consumption of 11.5 mpg. Install a 40 hp engine, as many builders do, and you get a 7.3 knot cruising speed and 7.3 mpg. Cruise that 40 hp engine at 6.6 knots and you are using only 12.5 hp, under a third of its rated power rather than the minimum recommended 50% (that's power, not rpm). The argument about having extra power for heavy weather has a serious hole in it. If you bear away about 20 to 30 degrees from the direction of the waves, you'll go faster, use less fuel and have a much more comfortable ride!

Semi-displacement and planing hulls

These hulls need much more power than a displacement hull. Because of the demands that the engine should be as light and as compact as possible, these engines are normally turbo-charged and have electronic engine management. To save carrying redundant weight, these engines are normally cruised at about 300 rpm below maximum continuous rpm. Heavy weather will require a reduction of speed, so you don't need any extra power.

Hull design and desired cruising speed affects the power requirement and it's not easy to use any rule of thumb, as it is for a displacement hull. Once the hull gets beyond 'displacement speed' you'll almost certainly be using enough power to avoid problems caused by running a diesel at too low a power. If you are forced to slow to displacement speed and you've got two engines, shut one down if safe to do so.

GETTING THE POWER TO THE PROP

Shaft drive

Traditionally, the propeller, or screw, is mounted directly on a shaft extending aft from the engine's

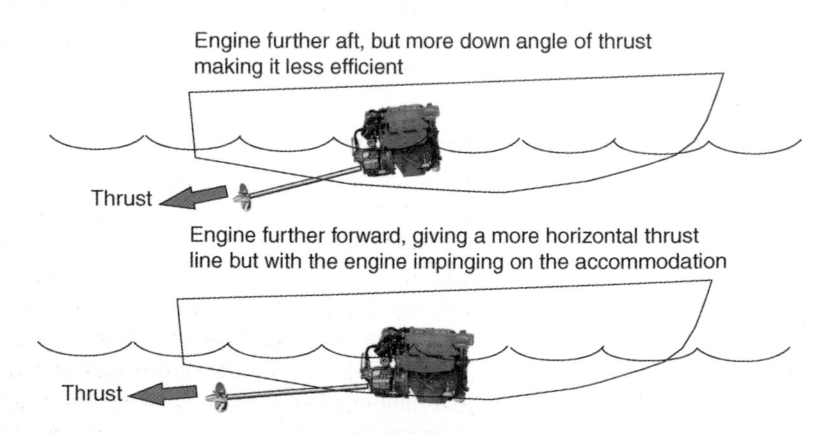

Conventional shaftdrive

'gearbox' and exiting through a waterproof gland towards the rear of the hull. Traditional hulls were relatively deep and the shaft could exit more or less horizontally. Modern hulls are relatively shallow, so if the downwards angle of the shaft is not to be too great, the engine needs to be mounted fairly well forward in the hull, but this may then intrude on the accommodation space.

Advantages:
- simple design;
- relatively cheap to make;
- easy maintenance;
- thrust bearings can be used so that no thrust load goes into the gearbox or through the engine mounts.

Reproduced by permission of John Bass

Disadvantages:
- engine and shaft need proper alignment if wear and vibration are to be reduced;
- the thrust line may be angled downwards.

An alternative solution is to use several shafts and angled gearboxes, either in the form of a 'Z' drive or a 'V' drive, to keep the engine further aft. This solution may help the weight distribution on some planing boats. 'Z' and 'V' drives are heavier and more costly than simple shaft drives.

Stern drive

Many planing motor cruisers have 'stern drives'. The engine is mounted right at the rear of the boat and drives the propeller through a stern drive leg and gearbox mounted to the rear of the boat's transom. The leg tilts to adjust the planing trim, and swivels to achieve steering. The boat has no rudder. If you like, it's a bit like an outboard engine but with the engine unit inside the boat. Driving a boat with a stern drive needs a different technique than that for a shaft drive.

Engine right aft, giving good accommodation, but suitable only for planing boats as engine weight is right aft

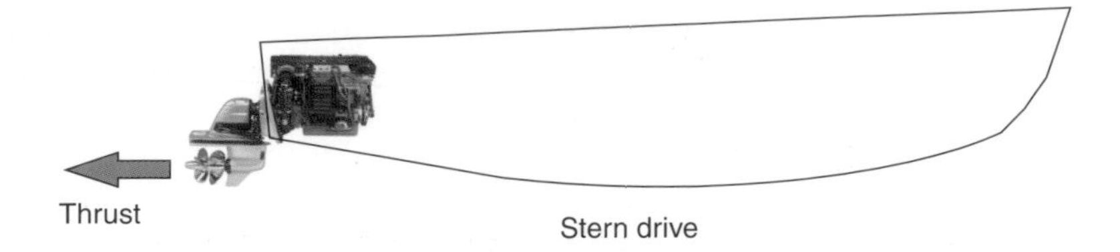

Advantages:

- engine weight can be kept far aft, an advantage in planing boats;
- installation costs are reduced with no engine alignment costs;
- with aft cockpit boats, engine accessibility is good;
- with aft cockpit boats, engines do not intrude into the accommodation;
- thrust angle can be 'trimmed' from a basic horizontal thrust line;
- speed is potentially greater than with a shaft drive.

Disadvantages:

- more expensive to build;
- externally mounted leg and drive unit needs frequent and expensive maintenance;
- electrolytic corrosion of 'out-drive' unit in salt water;
- boat has to be out of the water to service the gearbox/leg unit;
- with a deep 'V' hull configuration and twin engines, the engines have to be mounted so close together that servicing can be almost impossible.

Sail drives

A sail drive engine has its gearbox, leg and propeller all mounted as one unit, with the leg exiting through a hole in the bottom of the boat. This allows the engine to be mounted where it won't interfere with the

Normal arrangement

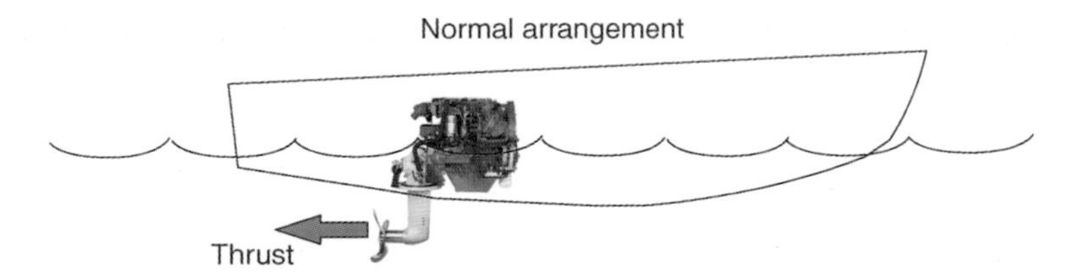

Engine reversed on leg giving more room for the accommodation but more weight aft

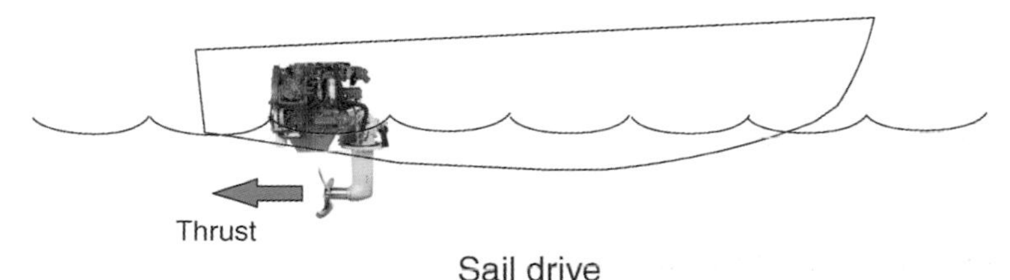

Sail drive

accommodation, but keeps the propeller's driving axis horizontal.

Advantages:

- installation costs are minimal;
- no engine alignment required;
- more choice of engine position, so its intrusion on the accommodation can be minimised;
- often less vibration (no shaft vibration as there would be in a poorly aligned shaft drive).

Disadvantages:

- large rubber diaphragm sealing hole in hull requires expensive replacement (every seven years for Volvo Penta, but not for Yanmar, which has a double diaphragm and a moisture detector);
- oil changes require the boat to be out of the water (except for the very latest Volvo and Yanmar models);
- possible corrosion of aluminium leg components in seawater;
- electrolytic corrosion of larger propellers as the leg anode is relatively small and often electrically isolated from the propeller;

- propeller mounted much further from the rudder, requiring more anticipation in close quarters manoeuvring;
- external water temperature may dictate non-optimal gearbox/leg lubricant;
- external water temperature may require non-standard battery charging until leg oil temperature has risen sufficiently to reduce friction drag;
- cannot use a thrust bearing, so all thrust is taken by the engine mounts and gearbox.

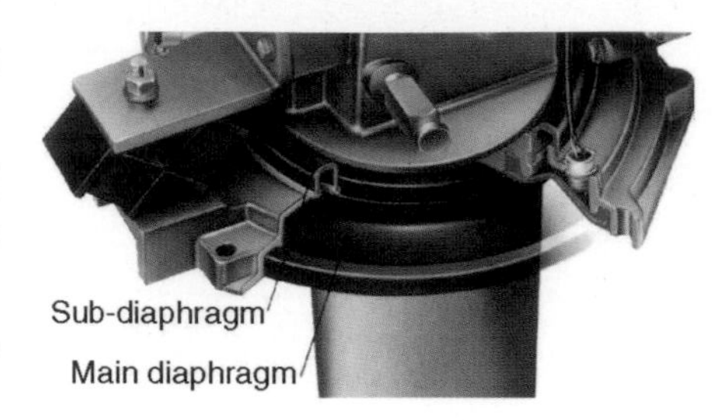

Volvo IPS

Introduced in 2004, Volvo's revolutionary IPS combines most of the advantages of the shaft and stern drives in one unit. It's a bit like a forward-facing sail drive with a steerable leg protruding from under the hull. The engine and drive are supplied complete and installed in pairs in fast motor cruisers.

IPS allows a horizontal thrust angle maximum efficiency and the engine far enough aft give good accommodation

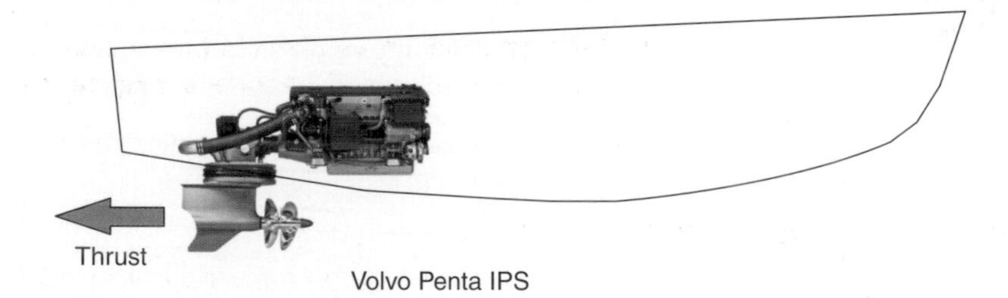

Volvo Penta IPS

Advantages:

- horizontal thrust line for higher speed potential;
- propeller in front of 'leg' in clear water;
- exhaust about 80 cm below the waterline to give very quiet running;
- steerable legs, giving good manoeuvrability;
- cast bronze under-water unit, giving good corrosion resistance;
- lower installation cost.

Disadvantages:

- high unit cost;
- available only with a couple of 4–500 hp Volvo Penta engines.

COMPRESSION IGNITION

A diesel engine has no ignition system or sparking plugs. Diesel fuel ignites at a temperature of around 320 Celsius, so what ignites the fuel and allows the engine to run? (*some writers give the ignition point as 900 °C. This arises from one document which translated °C to °F but then labelled the result in °C–most other writers followed suit!)* When air is compressed, work is done on the air, increasing its energy and thus its temperature. Provided that the air is compressed rapidly enough so that the heat has little time to escape to its surroundings, the air in a diesel engine cylinder can be made to rise to above the ignition temperature of the fuel by compression alone. If diesel fuel is then injected into the hot air, the mixture will ignite, releasing energy. This is known as *compression ignition*, unlike a petrol engine which uses *spark ignition* to ignite the fuel/air mixture.

Let's imagine an elephant jumping from a height onto a bag of cool air! And let's imagine that at the

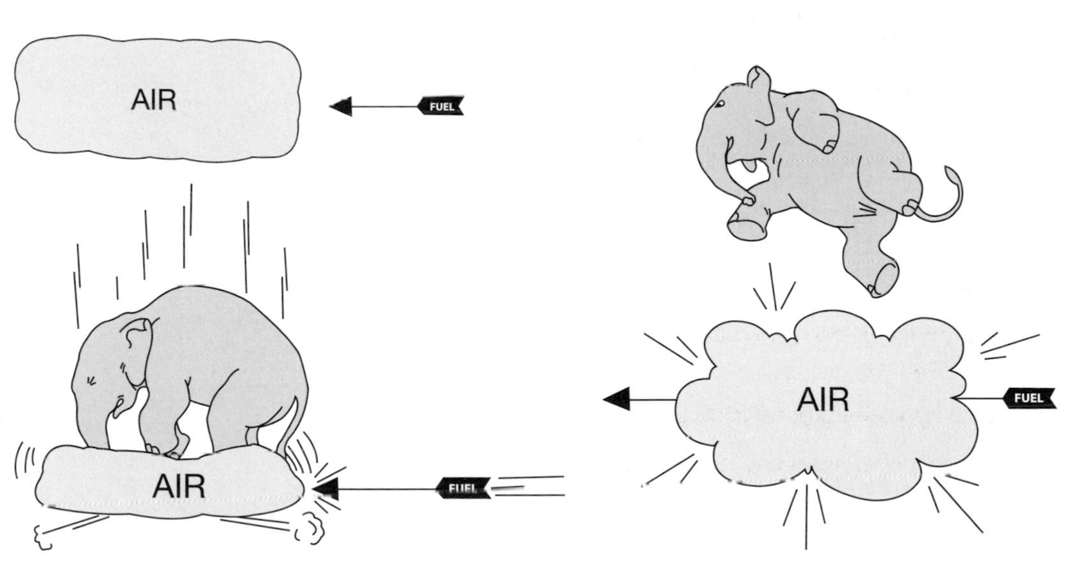

same time, an archer shoots an arrow full of diesel fuel aimed to arrive at the bag of air at exactly the same time as the elephant. As the bag of air is *very* rapidly compressed by the arrival of the elephant, the arrow with exactly the correct amount of fuel arrives and penetrates the bag of now very hot air. There's only one inevitable outcome: the elephant gets a free ride!

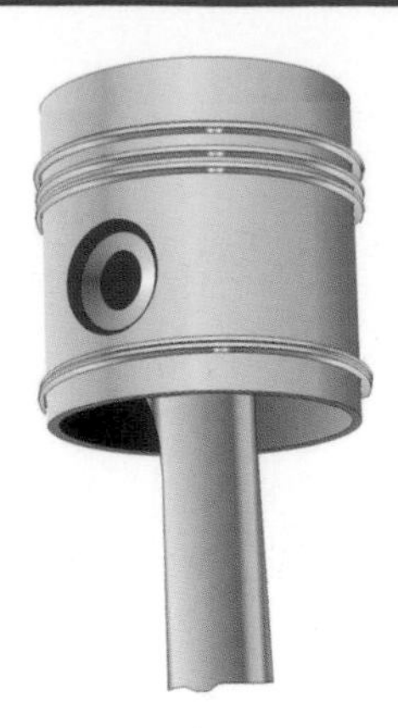

Very simplistic, I know, but the basic diesel engine is as simple as that. If the air is heated to above the combustion temperature of the fuel very rapidly AND if the correct amount of fuel is injected into this hot air at the correct time, the engine will run. No electricity is required, except to turn the engine over fast enough to start the engine.

Compression of the air takes place in the engine cylinder by reducing the volume of the air by around 20 times. In other words, there is a compression ratio of 20:1. The compression ratio by itself is of no use unless the air cannot leak out of the cylinder as the volume is reduced. Air is prevented from leaking past the piston as it moves in the cylinder by means of one or more piston rings, which press outwards against the cylinder wall to form a seal.

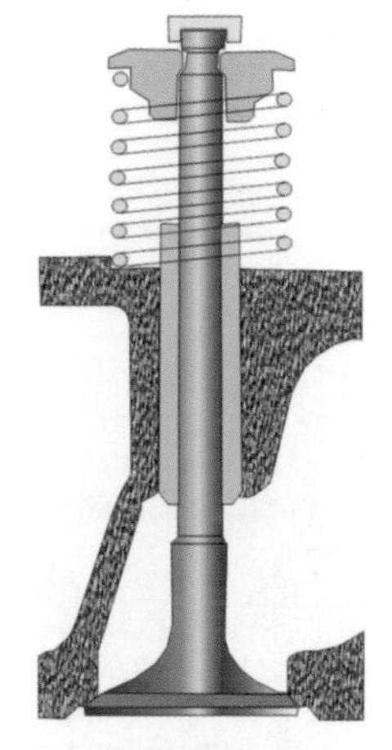

The inlet and exhaust valves must also seal properly on their seats. Valves must be seated properly to prevent gases escaping between the valve seat and the valve face.

THE FOUR STROKE CYCLE

Most diesels use the four-stroke cycle:

- Air is 'sucked' into the cylinder as the piston moves down with the inlet valve open – *the induction stroke*.
- Air is compressed and heated as the piston rises in the cylinder – *the compression stroke*.
- Fuel is injected into the hot air as the piston nears the top of its: compression stroke – ignition.

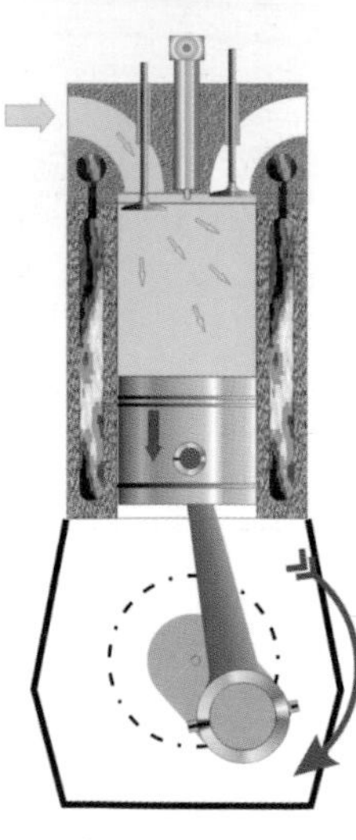

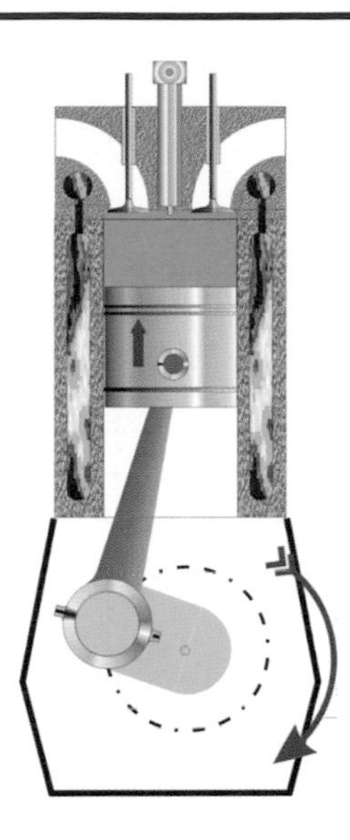

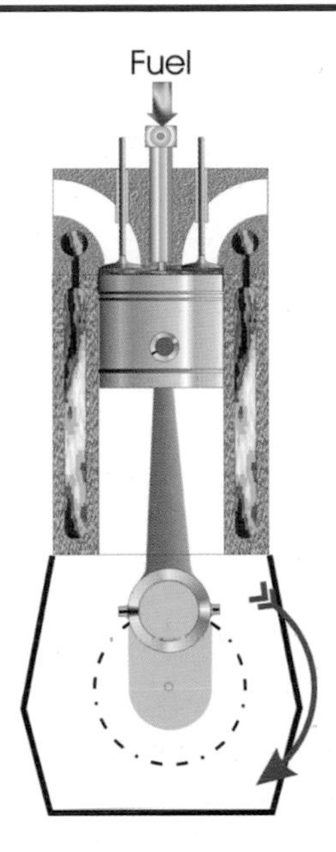

- Work is done on the piston by the rapid spontaneous combustion of the air and fuel mixture pushing the piston down – *the power stroke* .
- The piston rises, pushing the burned air/fuel mixture out of the cylinder with the exhaust valve open – *the exhaust stroke.*

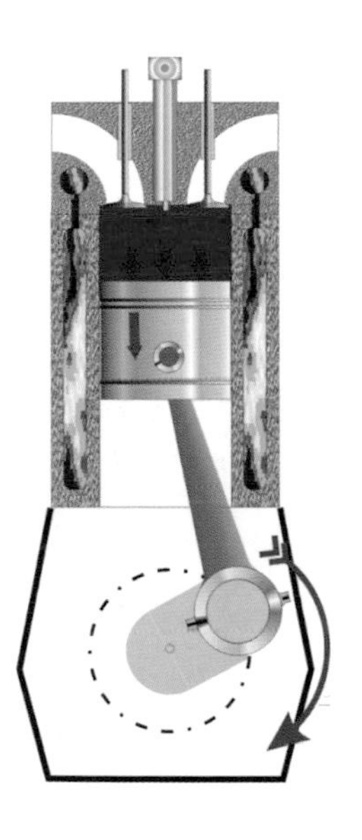

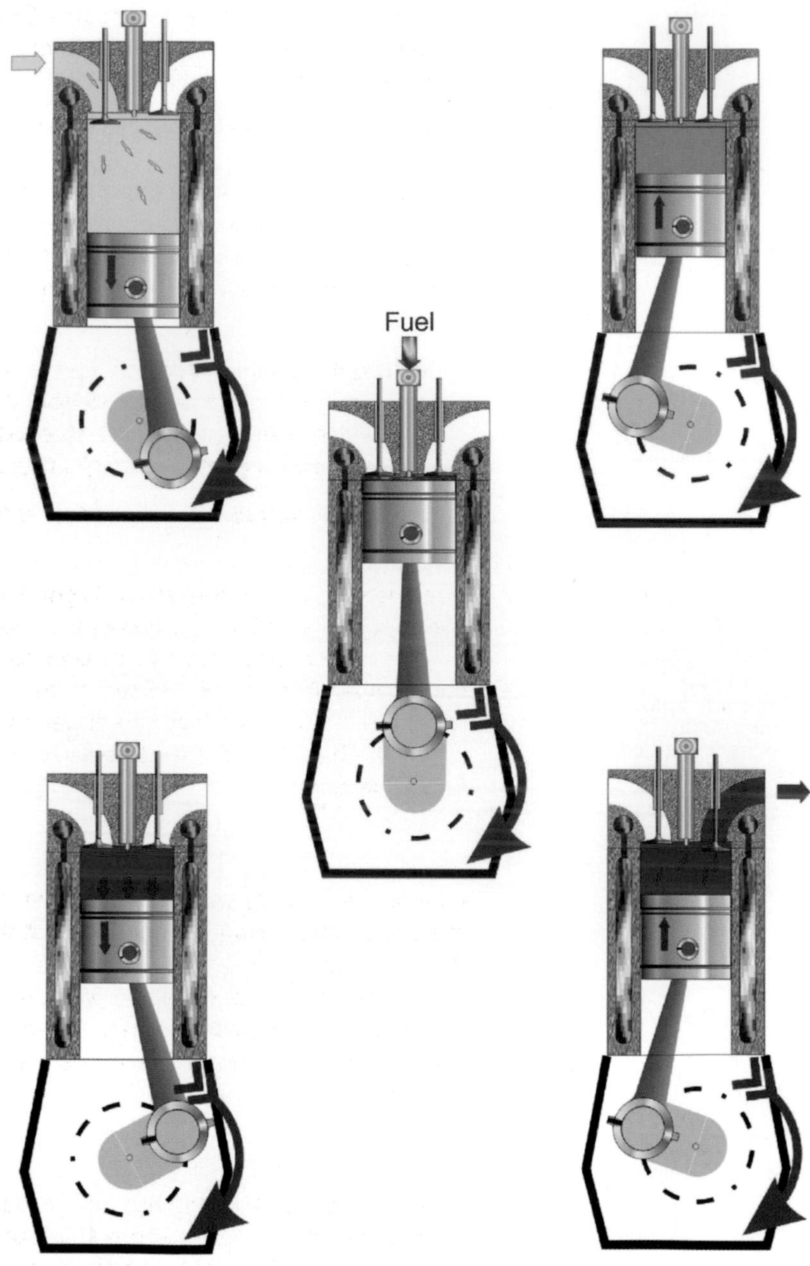

The four - stroke cycle

Note that the piston moves up and down twice for each 'bang' or power stroke. The valves operate only once for each two revolutions of the engine crankshaft. You get only one bang for each two revolutions in a single-cylinder, four-stroke engine.

The essential requirements for a diesel engine to start are:

- Adequate compression, supplied by the cylinder bore, piston rings and valve seats all being in excellent condition so that the temperature of the air is raised to the ignition temperature of the fuel.
- Rotating the engine quickly enough to obtain rapid compression to minimise the escape of heat from the cylinder – this requires a well-charged battery of sufficient power if electric starting is used.
- The correct quantity of fuel injected at the correct time.

It can be seen that a diesel engine in good mechanical condition *will* start if it is turned over rapidly enough to raise the air temperature to ignition point *and* the correct quantity of fuel is injected at the right time. It should be noted that other than powering the starter motor, the basic diesel engine requires no electricity for its operation.

MULTI-CYLINDER ENGINES

Four-stroke multi-cylinder engines have the crankshaft and valve timing arranged so that the 'bangs' don't all occur at the same time. The 'firing order' for the cylinders is designed to give the smoothest running. Some engines use gear-driven counterbalance weights in the crankcase to give even smoother running.

TURBO-CHARGING

There is a point at which adding more fuel to the air in the cylinder of a diesel engine will produce no extra power. The extra fuel will be wasted because there is insufficient air to burn it.

There are several ways to provide more air: make the engine physically bigger by having bigger cylinders or more cylinders of the same size, or feed the engine with heavier air. Making the engine bigger increases its weight as well as its size, and where we need lots of power from an engine, such as in a planing motor cruiser, this method will be impracticable. We need more power per kilogram of engine and more power per 'litre' of engine than this method can provide. However, we can 'feed' the engine with compressed air, and then add more fuel to this heavier air to produce more power. The normal method of achieving this is by adding a turbocharger to the engine.

There is still plenty of energy remaining in the exhaust gases that are ejected from the engine's cylinder after combustion. Putting a 'fan' in the exhaust gases can harness this energy. The fan drives a compressor, which then supplies the cylinder with 'heavy air' rather than air at atmospheric pressure. If you double the mass of air you supply, you can add twice the mass of fuel and theoretically produce twice the horsepower than you would get from a 'naturally aspirated' engine. Producing more power increases the load on the engine's components. Double the fuel and you double the 'bang' hitting the top of the piston. In order to retain reasonable reliability, the increase in power provided by turbocharging is usually limited to an increase of 50% of the normally aspirated engine. The diagram shows how its done.

Compressing the air, as we know, raises its temperature. Raising its temperature lowers its density. Thus, the increase in mass of the compressed air will not be as much as we thought, due to this increase in temperature. We will, therefore, need a bigger turbocharger to make up for this shortfall, and this will cost more money.

We could cool the air after it's been compressed to increase its density and allow us to use a smaller turbo-charger. This would give us a smaller and cheaper

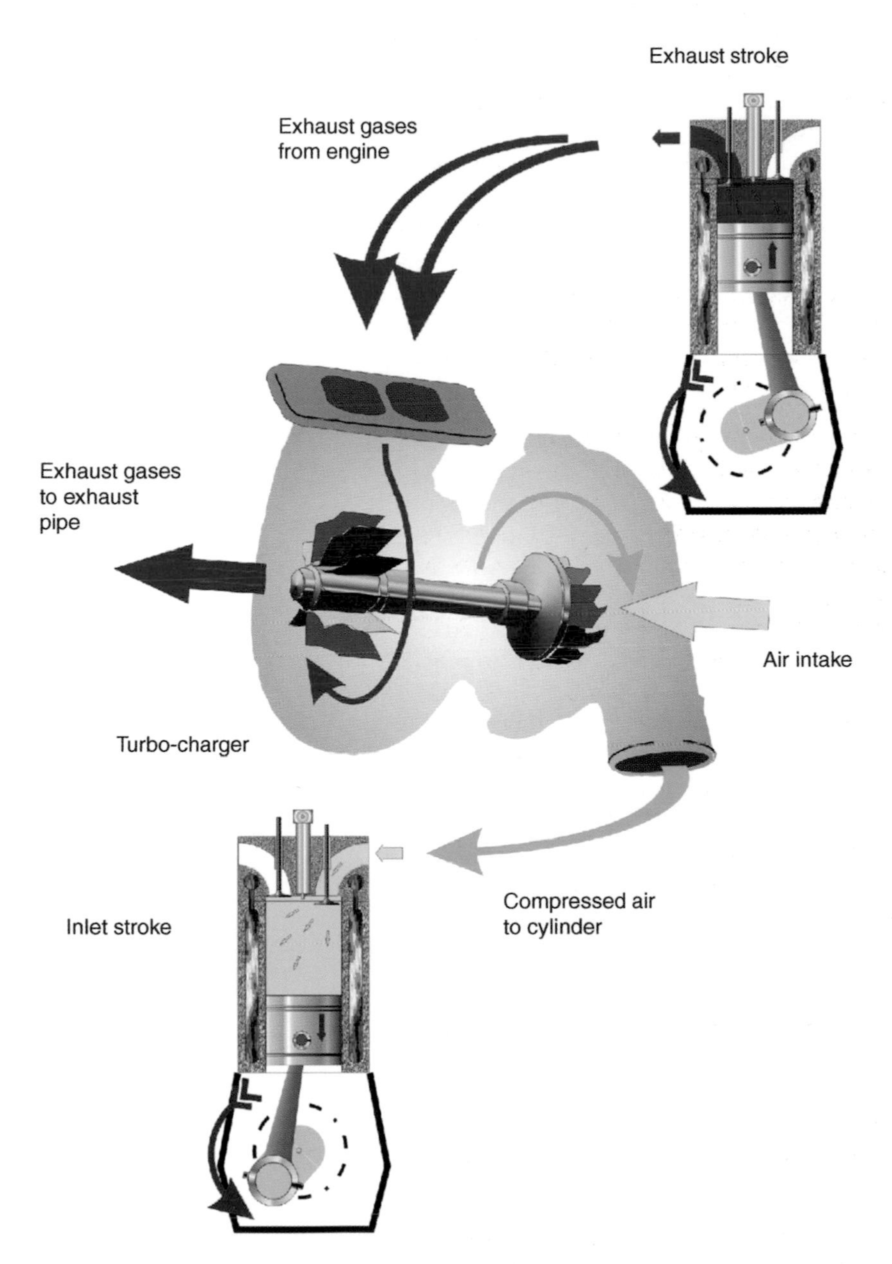

TURBO-CHARGING

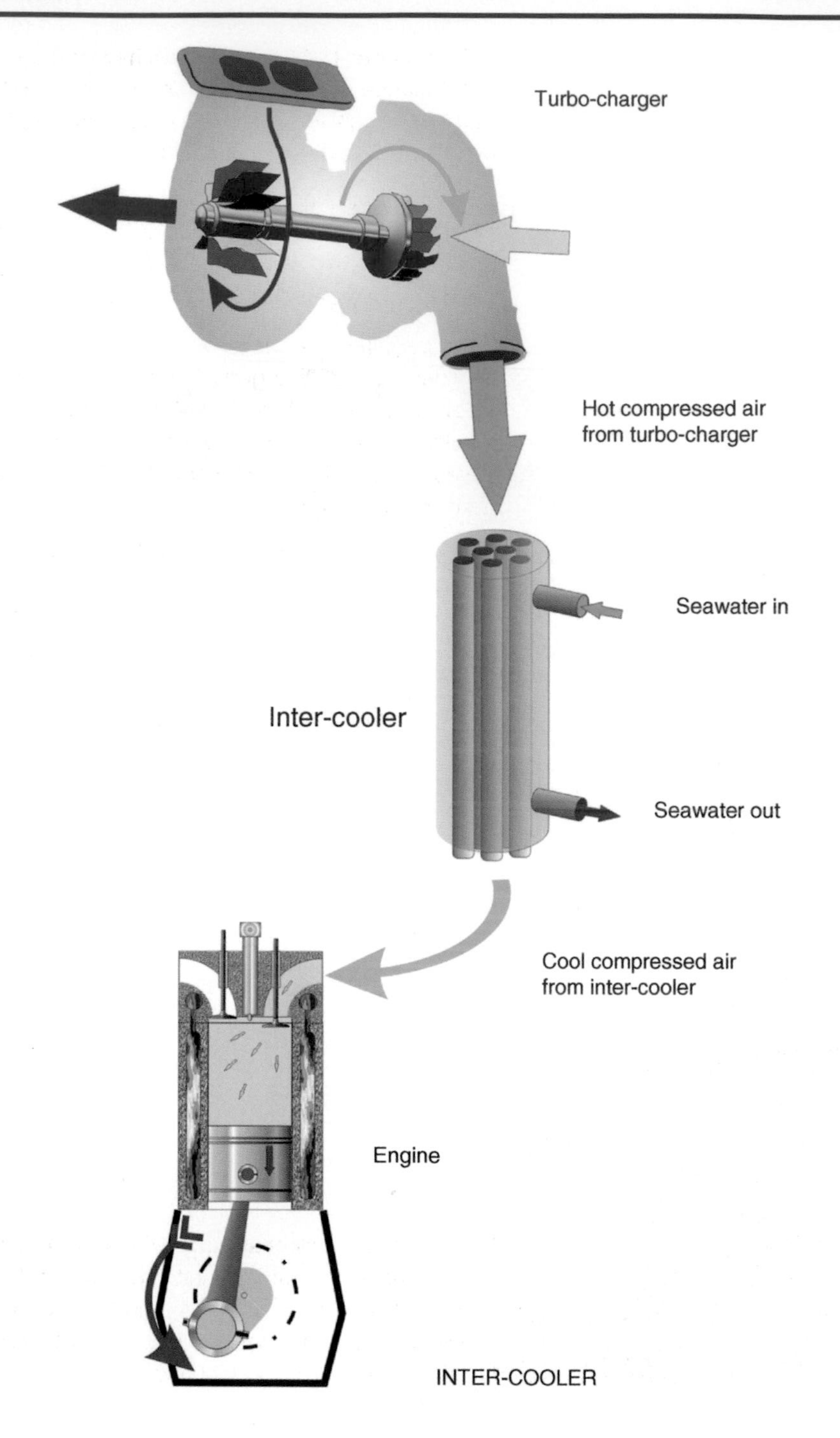

INTER-COOLER

turbo-charger, but we would now have to buy a cooler. The balance in cost occurs around 80 to 130 hp, so that smaller engines have no cooler and bigger ones do. It's all a matter of economics.

The cooler is fitted between the turbo-charger outlet and the engine air intake, and is usually called an *inter-cooler*. However, this same inter-cooler also goes by the name of *pre-cooler* or *after cooler*.

Road vehicles use air to cool the engine and turbo-charger heat exchangers. Marine engines use 'raw' water to cool the heat exchangers because there is insufficient air flow to give effective cooling in most cases. Raw water is the water the boat is floating in, normally seawater.

A turbo-charger adds power at higher rpm, rather than all across the rev range.

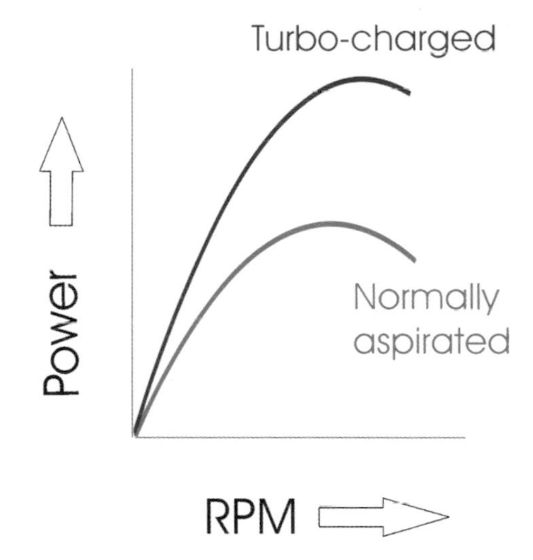

In order to add power all across the rev range, a bigger turbo-charger is needed, and this will give too much 'boost' at high rpm. To overcome this, an automatic *waste gate* is fitted, which allows excess pressure to be bled off to 'waste' at high rpm.

The life of a turbo-charger is a hard one. The turbine will have blade temperatures in excess of 900 °C, while just a few inches away, the compressor is running at around 100 °C. The whole rotor of compressor and turbine may be running at up to 120 000 rpm at maximum output.

There are certain things the boat owner can do to increase the longevity of a turbo-charger:

- make sure that the engine is idled for two minutes after running at high speed before stopping it;
- use only the recommended engine oil and change the oil at least as often as specified;
- make sure that the engine is not run at low power for any lengthy period of time to avoid carbon build up on the turbine blades.

This last requirement makes a turbo-charged engine a questionable choice for a sailing boat, which often is run at low power and then shut down before it is fully warmed up, or indeed an inland waterways boat, which is never run at much more than idle power at any time.

SUPER-CHARGING

A super-charger is a gear-driven compressor used to compress the air prior to delivery to the engine

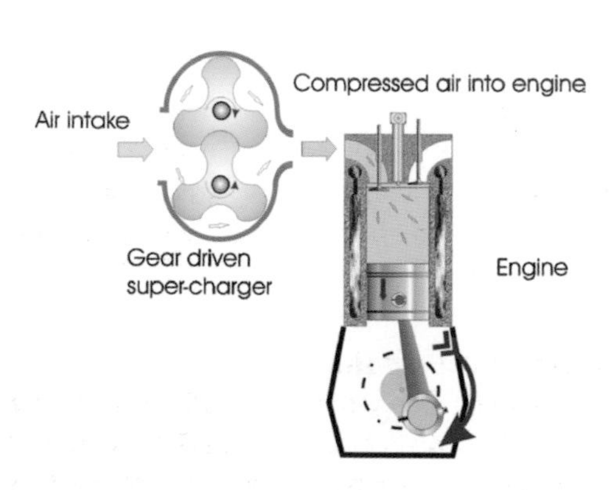

Super-charger

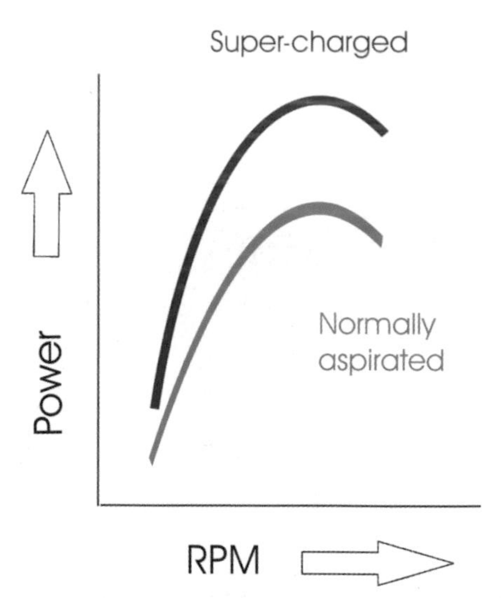

to increase the power of the basic engine in the same way as a turbo-charger. It can be geared to produce extra power at low rpm. It is not usually used on marine diesels. Volvo Penta does, however, use a combination of both super-charger *and* turbo-charger to increase the power over the whole rev range. This produces a relatively compact, high-power engine, at the expense of extra complexity.

THE TWO-STROKE DIESEL ENGINE

This is a very different beast from the four-stroke engine and also bears little resemblance to a two-stroke petrol engine, and there is only one player in the field – Detroit Diesels.

Because the four parts of the induction–compression–power–exhaust cycle are compressed into only one up and one down stroke in the two-stroke engine, it cannot compress the air enough to raise its temperature to the ignition point of diesel fuel unless there is some form of additional compression. The additional compression is provided by a gear-driven super-charger that compresses the air before it passes into the cylinder. Thus, the cylinder is supplied with pre-heated compressed air.

The smallest engine in the Detroit marine engine range is 400 hp.

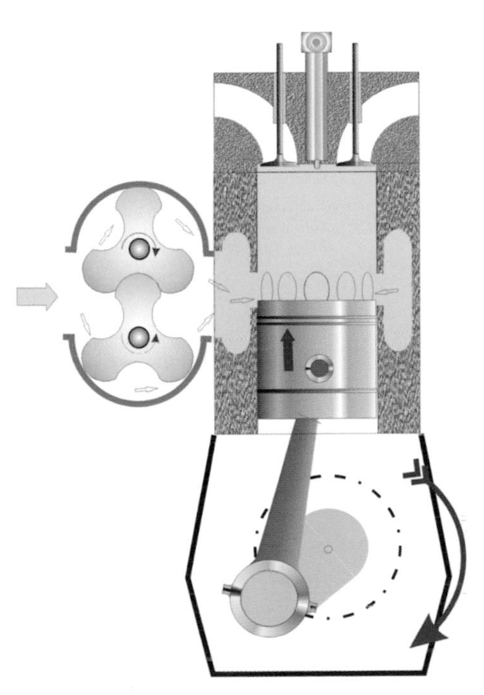

Inlet phase

At the lower portion of the piston's stroke, inlet ports in the side of the cylinder wall are uncovered, allowing compressed (and heated) air from the super-charger to enter the cylinder.

Compression phase

Once the inlet ports are covered by the upward moving piston, the air is compressed in the cylinder. Because the air has been compressed in the super-charger, by the time the piston reaches the top of its stroke, the air is above the ignition point of the fuel.

Ignition phase

With the piston approaching the top of its stroke, fuel is injected into the very hot air, where it ignites and burns very rapidly.

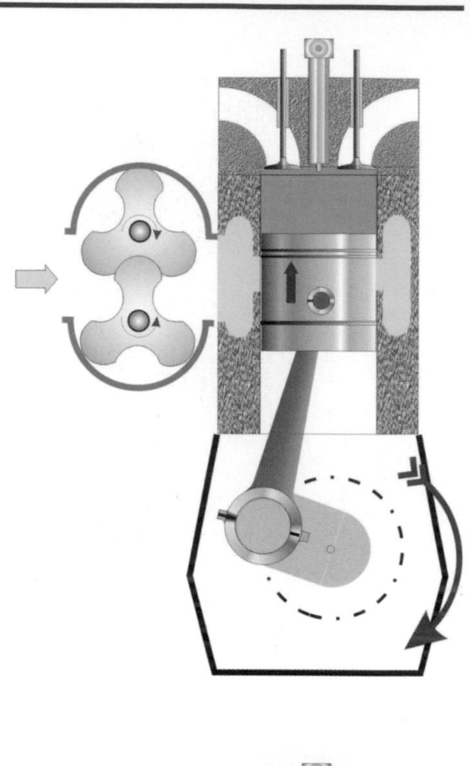

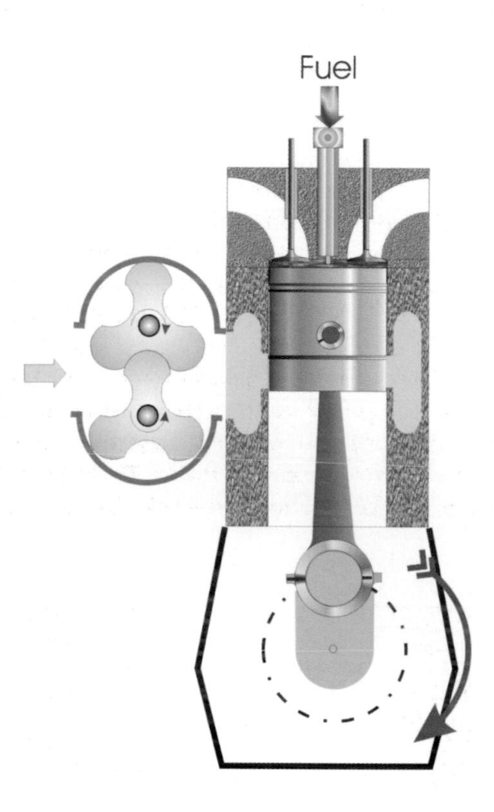

Power phase

The force created by the rapidly burning mixture pushes the piston downwards.

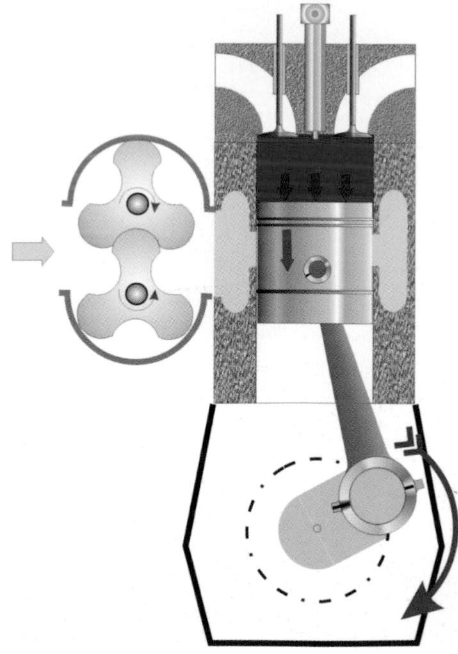

Exhaust phase

As the piston approaches the bottom of its stroke, the two exhaust valves open, allowing the exhaust gases to escape. At the same time, the inlet ports are uncovered, allowing compressed air into the cylinder. This

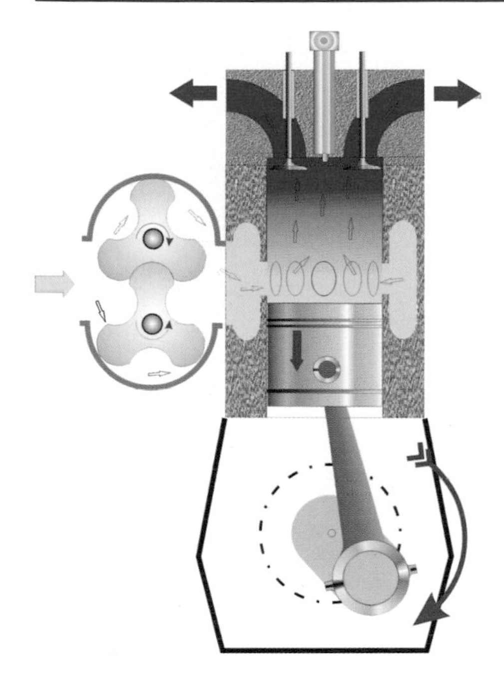

incoming air helps scavenge the exhaust gases from the cylinder.

The two-stroke diesel has a power 'stroke' once every revolution and so has twice as many 'bangs' as a four-stroke engine at the same rpm, so should give twice as much power as a similar sized four-stroke engine. However, its 'breathing' is not as efficient as a four-stroke, so, in reality, their power outputs are similar. Detroit Diesels claim very good fuel efficiency, but they now use them only for 'off-road' engines.

INJECTING THE FUEL

The fuel injection pump

The fuel system delivers fuel at low pressure to the fuel injection pump, which raises the pressure of the fuel to 2000 to 3000 psi (130 to 200 atmospheres) and meters out the fuel, as required, to the individual fuel injectors (atomisers). Mechanical injection pumps use individual pumps for each cylinder, although these pumps are usually part of a single injection pump unit mounted externally and gear driven. These have their governor (see p. 37) mounted inside the pump unit's body.

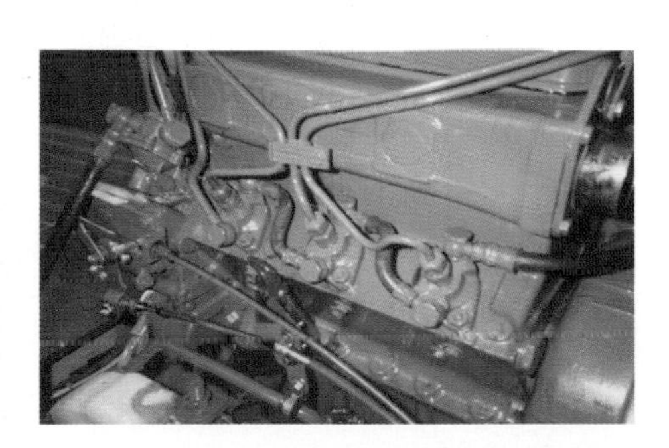

Smaller engines often have a set of pumps, mounted either collectively or individually, and operated directly by the engine's camshaft. In this case, the governor is part of the engine itself.

The amount of fuel delivered to each fuel injector is controlled by adjusting the effective stroke of the individual pumps by means of a fuel *rack*, whose position is controlled by the governor. Although the actual stroke of the pump plunger (piston) is constant and actuated by a cam, the position of the tapered groove in the plunger allows the volume of fuel delivered to be varied.

With the 'throttle' at idle, the rack is at its idle position and idle fuel is delivered to the fuel injector(s).

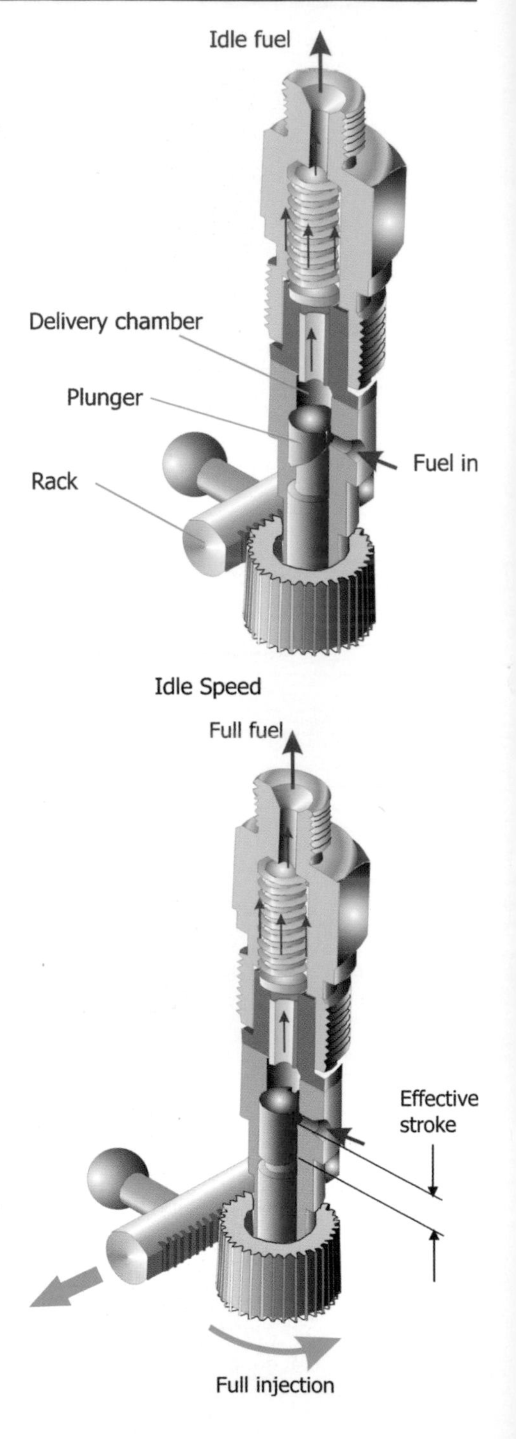

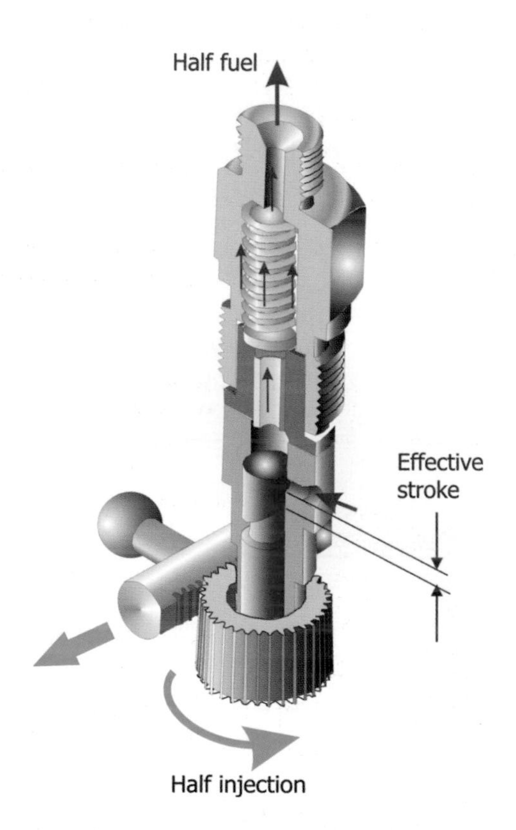

The fuel rack is moved to its half fuel position under the influence of the governor and rotates the gear and plunger so that the tapered groove gives more 'effective stroke' before it cuts off the fuel supply.

With the rack at full travel, the tapered groove allows 'full stroke' and delivers maximum fuel.

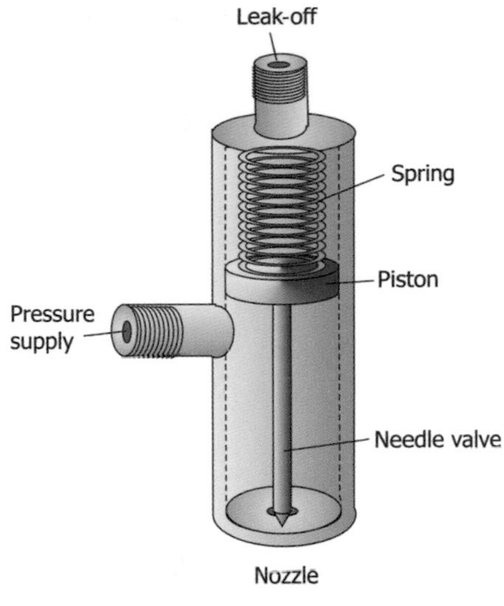

Fuel Injector

Some engines use a slightly different type of injection pump and have a hydraulic governor. Essentially, all fuel injection pumps are precision-made and cannot be serviced by the user.

The injector

The pressurised pulse of fuel from the injection pumps doesn't reach full pressure instantaneously. If the injector consisted of one simple orifice, fuel would dribble from it into the combustion chamber until the pressure rose sufficiently to form a spray. The dribbled fuel wouldn't burn, so would be wasted and would produce smoke. This is undesirable, so the injector has a valve mechanism within.

1. A spring, bearing against a piston, holds a needle valve closed against the orifice.
2. As the fuel pressure rises, the piston moves upwards, lifting the needle valve and opening the

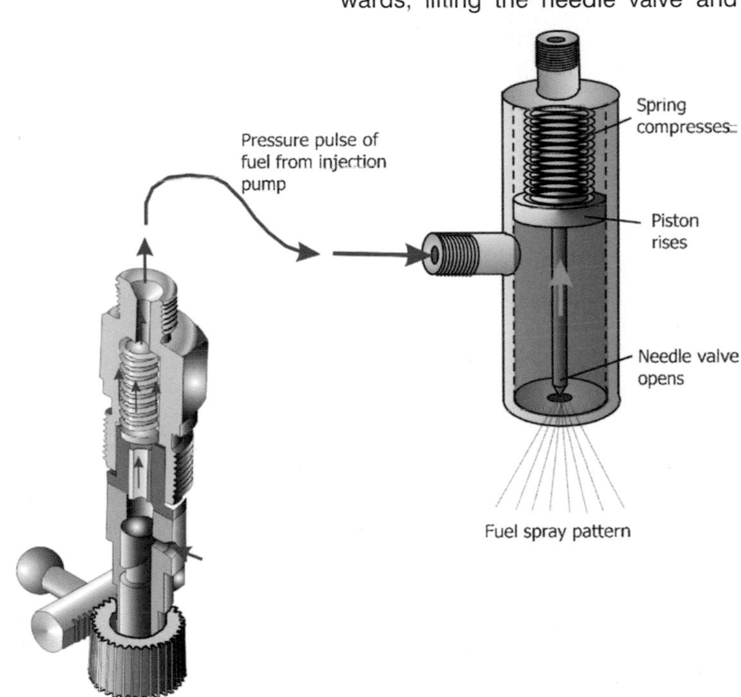

Injector valve open

orifice, allowing a controlled spray pattern to emerge.

3. The piston has a simple seal inside the injector and high pressure fuel will leak past it. This fuel would fill the cavity above the piston, preventing the piston from rising on the next pulse. To prevent this, a leak-off pipe ensures that only very low pressure exists above the piston. The leak-off fuel is returned to the fuel tank or to the fuel filter inlet.

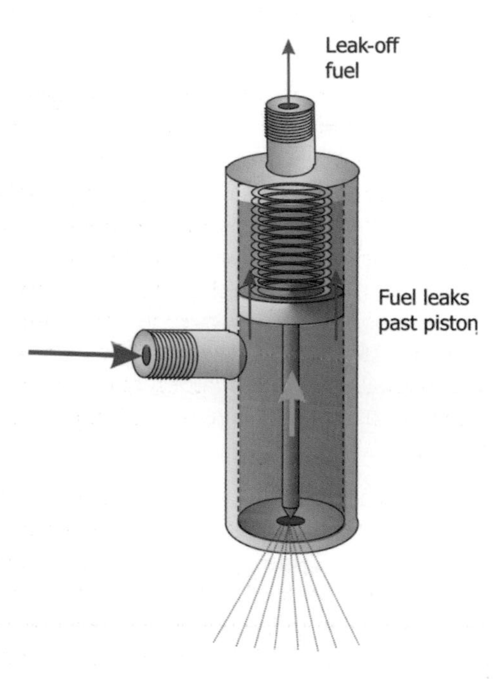

Injector leak-off

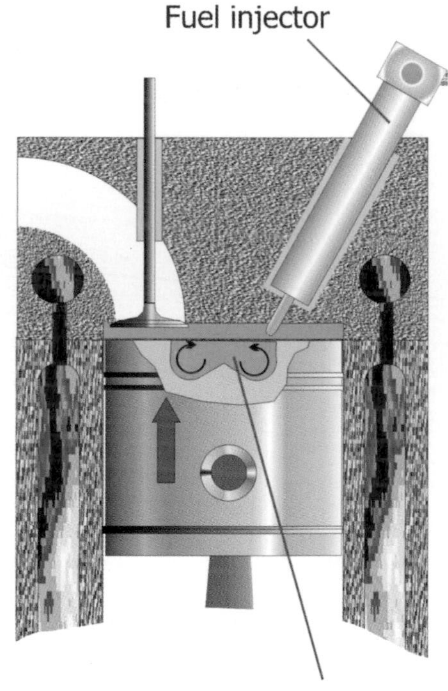

Direct injection

Direct injection

For many years, fuel was injected directly into the cylinder. The top of the piston is not flat, but contains a large depression that acts as the combustion chamber. The shape of the depression is such that it tends to induce a swirling motion to the air as the piston rises

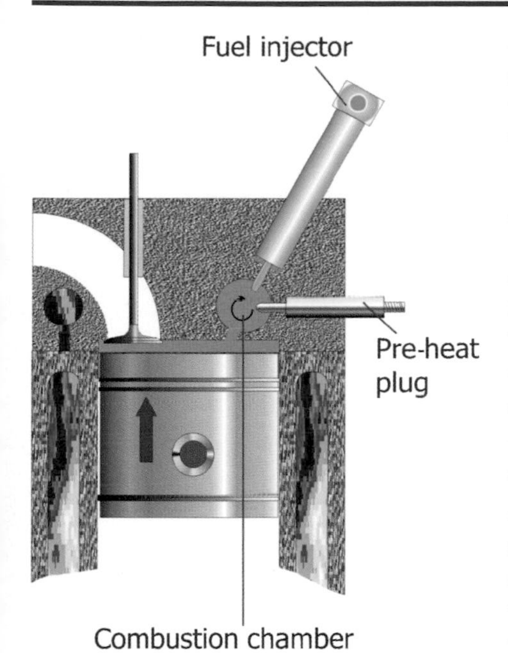

Indirect injection

on the compression stroke. The injector directs the fuel spray, at an angle, towards the depression, so that the fuel is mixed with the swirling air in as homogenous a way as possible.

Indirect injection

In order to give better mixing of fuel and air to achieve more efficient combustion and less emissions, a move to indirect injection was made. A further gain was smoother running. In this set-up, the combustion chamber is situated outside the cylinder and the fuel is injected into this external combustion chamber.

The combustion chamber can be designed to give much better mixing of the fuel and air and is sometimes known as a *swirl* chamber. The very small cross-sectional area of the passage linking the cylinder and the combustion chamber reduces the shock of the combustion 'bang' hitting the piston and increases the time that the 'bang' pushes down on the piston. This makes for smoother running. Despite the better combustion efficiency and reduced emissions, the restriction reduces overall power and so an overall increase in fuel consumption occurs.

Indirect combustion engines will not start from cold at low temperatures. Not all the potential increase in temperature occurring during compression is available, as the extra metal surrounding the combustion chamber 'robs' the heat from the hot air. Indirectly injected engines need some form of *pre-heat* to achieve cold starting.

Many modern designs of diesel engine are returning to direct injection because electronic fuel injection can reduce emissions without recourse to indirect injection with its attendant power loss.

Common rail injection

Modern car engines often have common rail injection. The high-pressure fuel pump doesn't measure the fuel, but supplies the fuel at injection pressure to a storage manifold running along the top of the engine. The fuel injectors, which are controlled electronically and fed directly from the manifold, open as required to allow a measured quantity of fuel into the cylinder. This system achieves improved efficiency and emission control, and is beginning to find its way into marine engines.

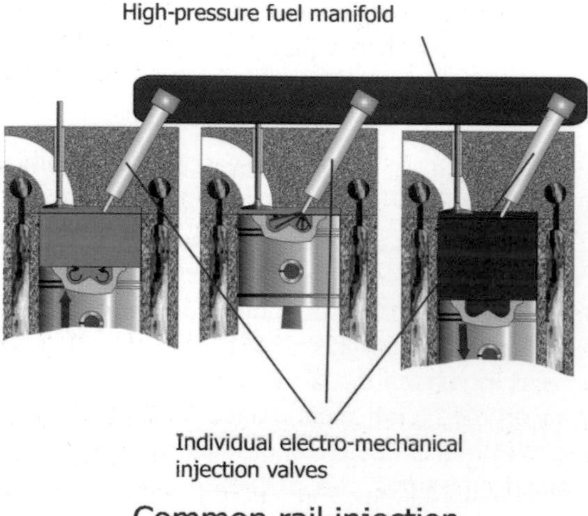

Common rail injection

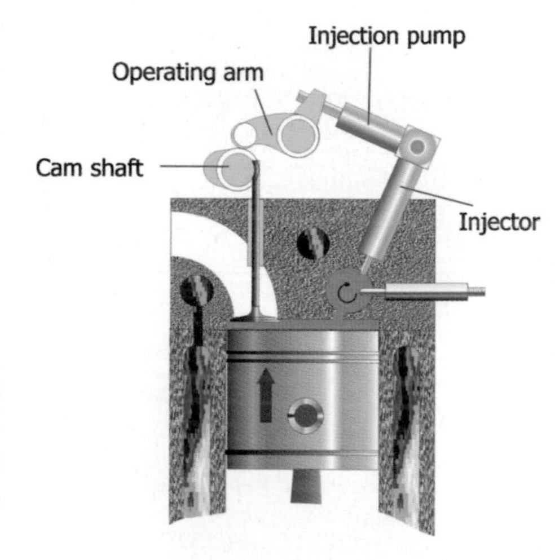

Combined injection pump and injector

Some engines, such as Lombardini, use individual combined fuel injection pumps and fuel injectors operated directly by the overhead valve camshaft. They are supplied from a low-pressure fuel supply and again have no central fuel injection pump.

COLD STARTING

When it comes to cold starting, there's just no substitute for reading the handbook! You would be amazed at how many people who have owned their boat for some time have no idea how to start the engine from cold.

Extra fuel for starting

Many direct injection engines use extra fuel for cold starting. On some, like the small Yanmars, this extra

fuel is supplied automatically, and the owner need do nothing except set the 'throttle' correctly. On others, however, special actions have to be taken prior to starting from cold. For instance, on the old Volvo Penta MDI, 2 and 3 and their derivatives, there's a cold start button at the back of the engine that must be pressed down for each cold start attempt. This action allows the fuel control rack to run to a 'cold start' position by removing a rack normal stop. This is often situated under the cockpit sole and many owners are unaware of its existence. In some cases, an enterprising owner has rigged up a cable-operated remote control for it.

The Volvo Penta MD 2000 series engines require that the 'throttle' should be opened about two-thirds (the book says less, but in my experience you need more) and *then*, afterwards, pull the STOP handle and return it to run before starting. If the engine fires but doesn't run, you must do this again. The rack stop is temporarily removed as for the case above.

On the Petter mini 6 and mini twin, a wire loop will be found on the front of the engine. This is pushed down and released after the 'throttle' has been set and before starting.

Raising the compression ratio

The engine's compression ratio can be increased for cold starting on some engines by the addition of lubricating oil to the cylinder above the piston. In particular, this is done on the Petter mini 6 and the mini twin and the Sabb, as illustrated.

Pre-heating

Engines having indirect injection are fitted with pre-heating as an aid to cold starting, and some direct engines may be fitted with some form of pre-heating as an option for use in low-temperature situations.

Pre-heater plugs

Electric pre-heater plugs (glow plugs) are fitted permanently in the combustion chamber and powered prior to start from the engine's electrical system. The heater pre-heats the whole combustion chamber. Its tip, which extends into the combustion chamber, is still hot when the fuel is injected, aiding ignition.

Where used on cars, an orange light on the instrument panel is illuminated when pre-heat is selected and goes out when the required temperature is reached, indicating the starter motor may now be engaged. On most marine engines, this indication is not given, so the operator needs to be aware how long pre-heat should be applied for. If the battery is so low that engine starting may be impaired, correct use of pre-heating is essential – read the manual.

Other forms of pre-heat

There are other forms of pre-heat on some engines.

Thermostart

A fuel reservoir supplies a burner in the air intake manifold. This burner is ignited during a cold start to preheat the air and was standard on Perkins 4-107, 4-108 and 4-109 engines.

Chemical igniters

Sabb engines use a chemical 'cigarette' inserted into a special holder, which is ignited by the elevated compression pressure during starting. This raises the

temperature further to achieve combustion temperature. These are used only for very low-temperature starting, not as a matter of course.

Electric heater in the air intake
An electrical coil, mounted in the air intake manifold, is supplied with electric power during cold starting to pre-heat the air as it passes to the cylinder.

VALVE GEAR

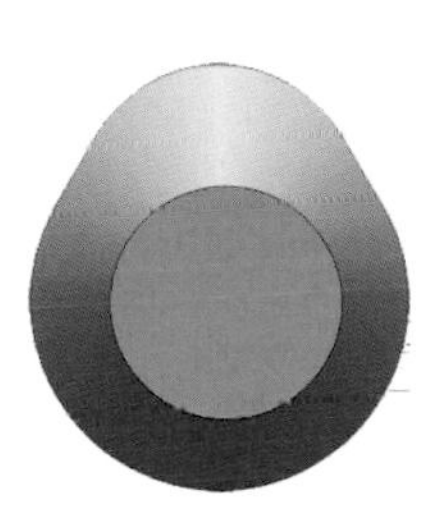

The inlet and exhaust valves have to be opened and closed once every two revolutions of the crankshaft. They are operated by a shaft (the camshaft) running at half the rotational speed of the engine.

A cam is a circle with a 'bump' on it. As the bump rotates, it is able to operate a mechanical device. A number of cams are machined onto a *camshaft*.

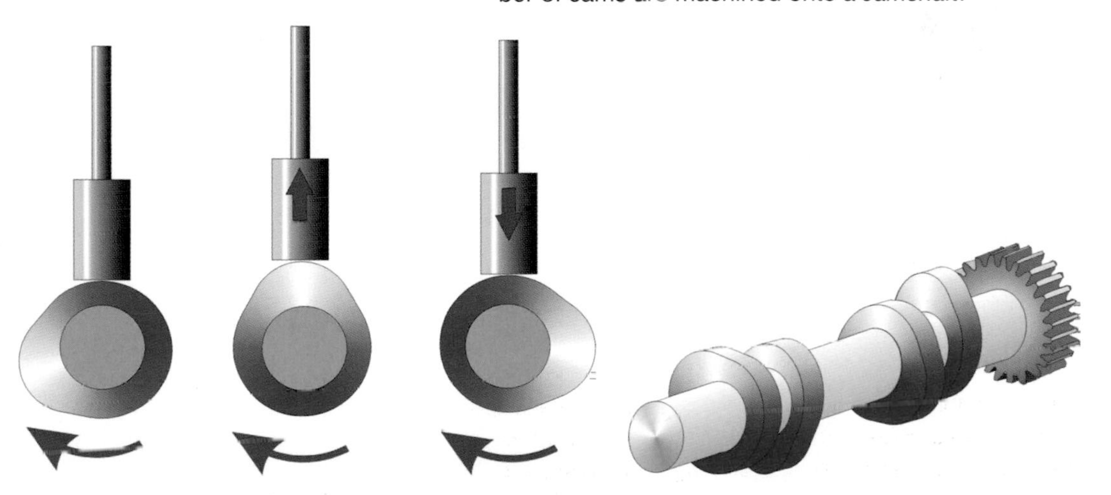

Push-rod-operated valves

The camshaft is mounted low in the engine and is gear driven. Long push rods operate the overhead valve gear. This is a cheap and efficient option, but the inertia of the push rods make it unsuited to high-speed engines. Where the engine is based on an industrial engine or is a dedicated marine engine, this will be the norm. Adjustment of valve clearance is easy, (see Chapter 14).

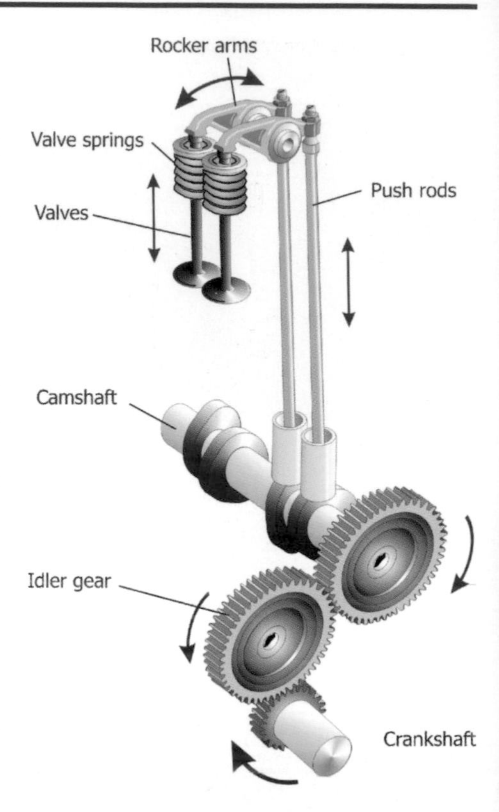

Overhead camshaft-operated valves

Engines designed for high rpm will be fitted with one or more overhead camshafts. These engines will be derived from car diesel engines. The overhead camshaft will be driven by a gear train, chain and sprocket or by a rubber cam-belt.

Gear-driven overhead camshaft

This is reliable, heavy and expensive. It is the ideal.

Chain-driven camshaft

This is lighter and cheaper than using gears. It uses engine oil pressure to operate the automatic

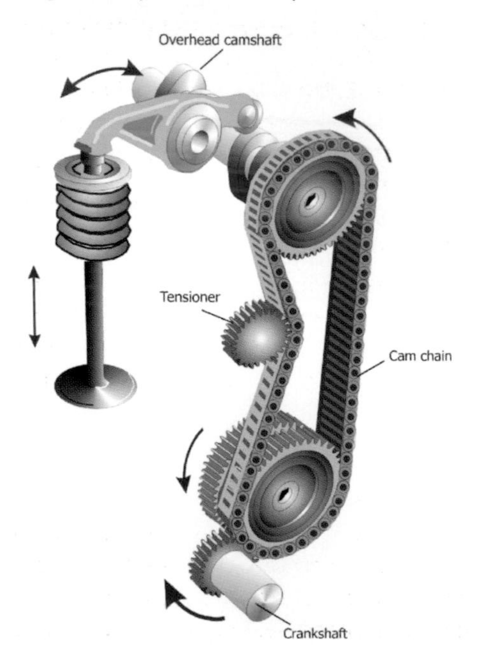

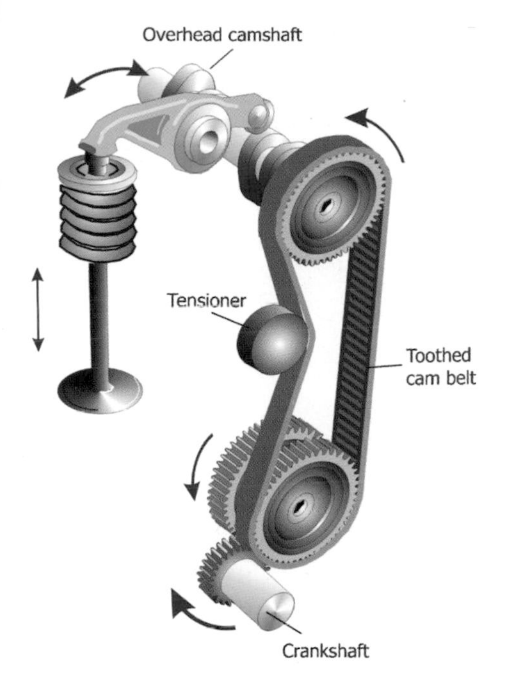

chain tensioner and, provided that the engine oil is changed at least as regularly as specified, it is reliable.

Rubber cam-belt drive

The cheapest and least reliable option is the use of a rubber cam-belt. This has long been popular in the car industry but requires the cam-belt to be changed at regular intervals if hugely expensive engine repair bills are to be avoided. Independent car experts agree that the cam-belt should be changed at 40 000 mile intervals, whatever the manufacturer says, and that the tensioner, if plastic, should be changed at the same time. The Ford Escort engine (which forms the basis of some marinised engines) requires a cam-belt change at 30 000 mile intervals. (40 000 road miles equates to around 800 engine hours at sea.)

How do we know if we have a rubber cam-belt? Look at the front of the engine. If the camshaft drive cover is held in place by easily removed clips or screws, your engine has a cam-belt.

Marinised engines that have a rubber cam-belt include the Volvo Penta MD22, Ford 1.5 and 1.8 derived engines and VW Golf-derived engines. Some Vetus engines may also incorporate rubber cam-belts.

If you have an engine with a cam-belt and its history is unknown, you would be well advised to have the belt changed.

In order to meet future emission controls, some car manufacturers are returning to chain-driven camshafts because the slight stretching of the rubber cam-belt gives less precision to the valve and ignition timing.

DECOMPRESSORS

Older non-automotive engines have decompressors. These allow the engine to be turned over with no compression as an aid to starting or for maintenance by partially opening the exhaust valve. It's essential that this is adjusted so that there's no contact between the exhaust valve stem and the decompressor arm when in the 'parked' position to avoid damage.

You can't hand start a diesel against compression, but if the decompressor(s) is/are used, the engine can be rotated by hand up to a speed at which it will start. Closing the decompressors will then allow the engine to start. Some multi-cylinder engines allow cylinders to be decompressed individually. When up to speed, close only one decompressor until the engine starts on one cylinder, then close the rest.

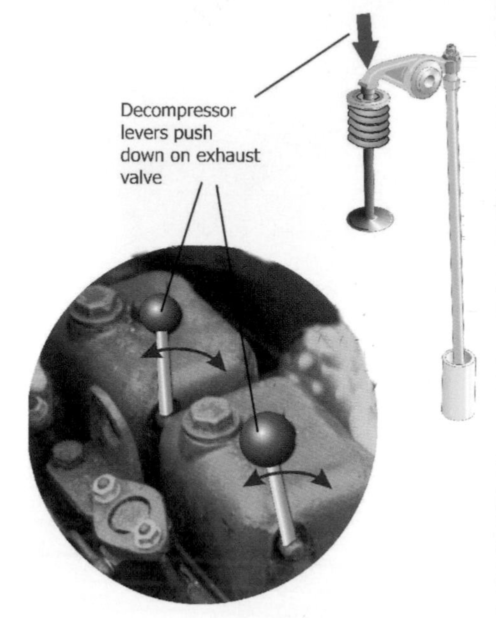

With a low battery, starting with the decompressors open will allow the depleted battery to turn the engine fast enough to start, the decompressors can then be closed.

Note that although the Volvo Penta 2000 series engines have only one decompression lever, in the fully up position, all cylinders are decompressed. With the lever moved to the left, only one cylinder is compressed to allow starting on one cylinder.

DIESEL ENGINE SPEED CONTROL

Unlike a petrol engine, a diesel engine has no throttle. A petrol engine uses a valve in the air intake manifold to control the amount of air entering the cylinder. Fuel is added to this controlled volume of air to give a constant air/fuel ratio of around 17:1 for normal running at constant speed. For increased power, more air is sent to the cylinder by 'opening the throttle', requiring more fuel, which in turn delivers more power.

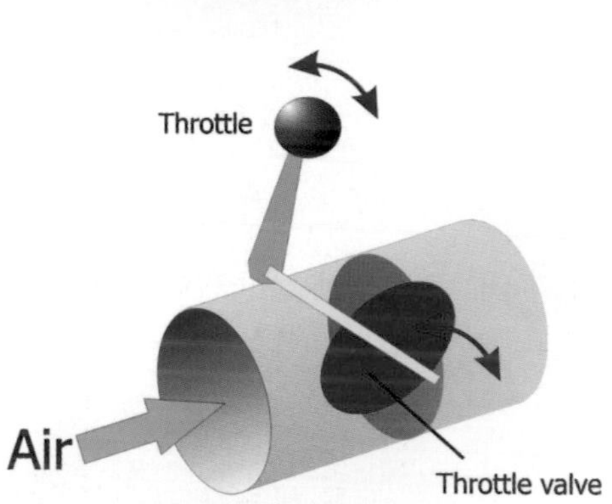

Throttle valve position determines how much air (and hence fuel) enters the cylinder of a petrol engine

If we were to restrict the air entering the cylinder of a diesel engine, compressing the restricted amount of air would fail to achieve combustion temperature and the engine would not run. A DIESEL ENGINE CANNOT HAVE A THROTTLE. Every induction stroke 'sucks' in a full cylinder of air.

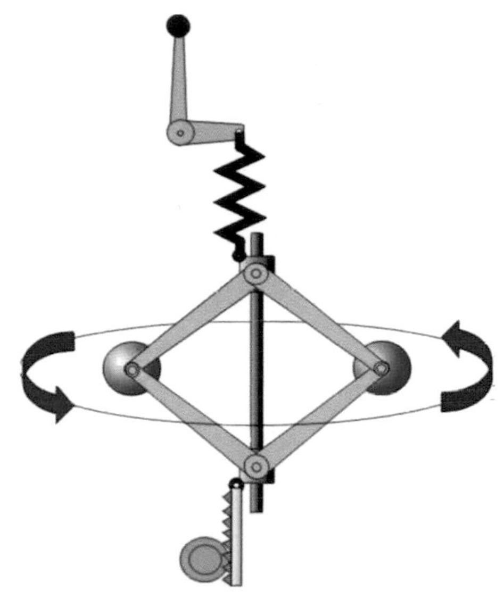

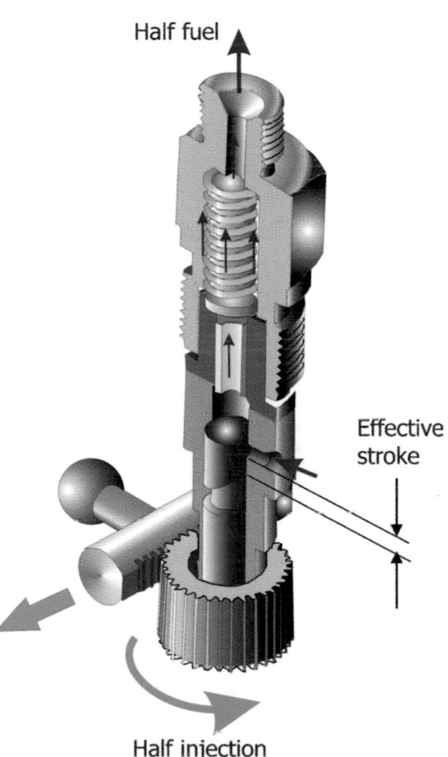

In a diesel engine, power output is controlled in a roundabout way. The engine speed is controlled by the speed lever ('throttle', gas pedal, or whatever you wish to call it). Just sufficient fuel is added to overcome the load on the engine to maintain the requested engine speed (rpm). This is achieved by a speed *governor* (regulator) working in conjunction with the fuel injection pump.

The governor

- A set of weights is spun round, driven by the crankshaft.
- The faster these weights are spun round, the further they want to fly out under the influence of centripetal or centrifugal force.
- Springs control how far these weights fly out.
- The tension of these springs is adjusted by the speed control lever ('throttle').
- The force on the springs is balanced by the centripetal force on the weights to control the speed of the engine.
- Any movement of the weights inwards or outwards forces a fuel control 'rack' to move.
- Movement of the rack causes a fuel control 'pinion' to rotate.
- The angular position of the pinion shaft determines the quantity of fuel delivered by the fuel injection pump.
- If the rpm falls, the springs pull the weights inwards.
- This inward movement signals the fuel injection pump to supply more fuel.
- If the speed rises, the centripetal force pulls the weights outwards.
- This outward movement signals the fuel injection pump to reduce the supply of fuel.

- The system is in 'balance' when the outward force of the weights is balanced by the inward pull of the springs.
- The engine rpm is then that which has been requested by the speed lever.
- *The speed lever determines the speed that the engine should run.*
- The fuel injection pump delivers just sufficient fuel to overcome the load on the engine to maintain the requested rpm.

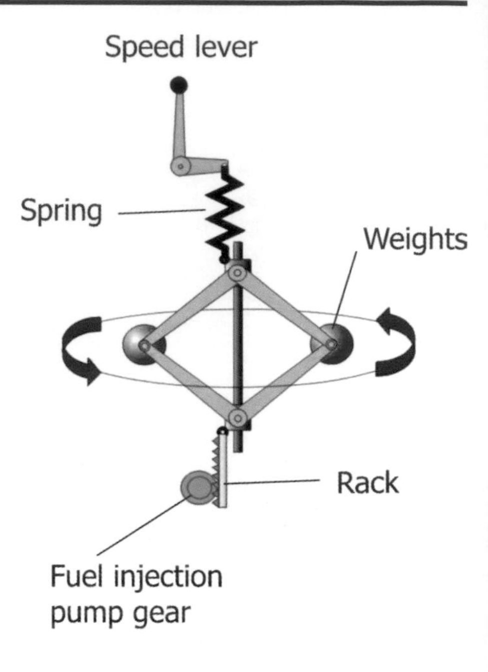

Idle RPM

Accelerating the engine

- To signal a request to increase engine speed, say from idle to 2500 rpm, the speed lever is moved ('opened') to the approximate position to give the required engine rpm.
- This opens the rack to the 'maximum fuel' position to accelerate the engine.

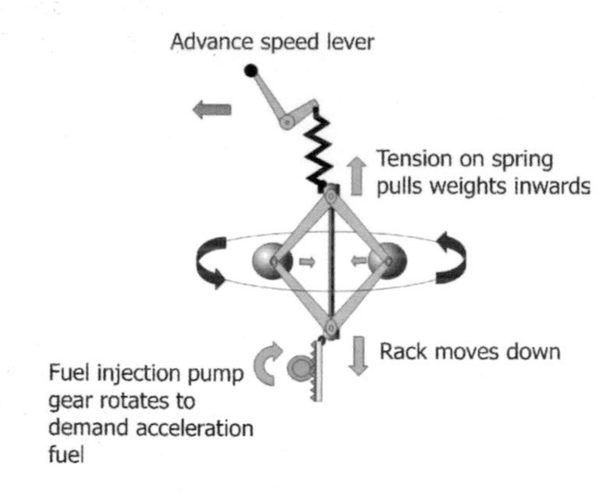

Acceleration

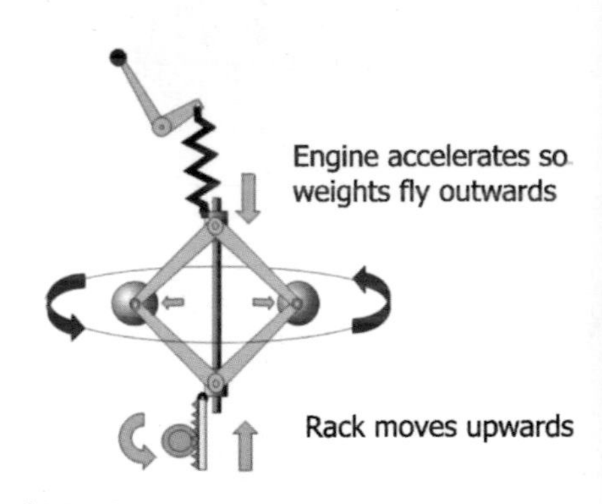

Steady RPM

- As the engine rpm approaches the requested rpm, the governor takes over and controls the fuel input to balance the load on the engine.

Slowing the engine

- The speed lever is 'closed' to the idle position.
- This moves the rack to the idle position.

- Pre-set idle fuel is supplied to the engine.
- The engine slows under the influence of the load to idle rpm.

Too much load

If the load is so great that the requested rpm cannot be achieved, the governor will not control engine speed and acceleration fuel will continue to be supplied to the engine, even though the engine is running at a constant (too low) rpm.

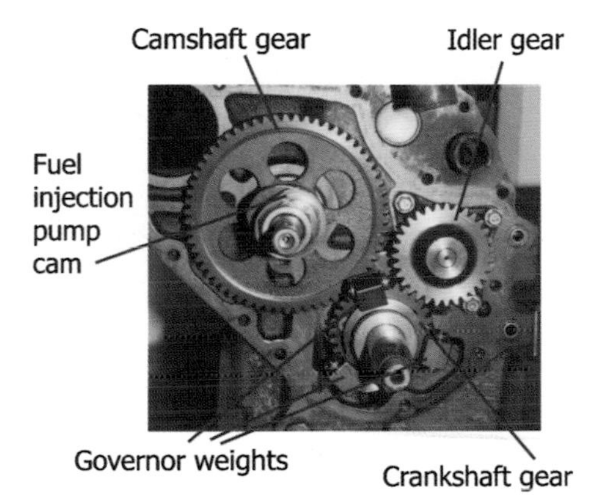

Where is the governor?

If the fuel injection pump is mounted on the engine casing and driven directly by the engine camshaft, the governor is inside the timing gear-case of the engine.

If the fuel injection pump is a separate unit, mounted externally to the engine, the governor will be mounted inside the pump casing. Some of these fuel injection pumps may have a hydraulically controlled governor.

Engines with electronically controlled fuel injection will have electronic governors.

SMOKE IN THE EXHAUST

Once the engine has reached normal operating temperature, its exhaust should be free of visible smoke. However, some engines, such as the Volvo Penta 2000 series, always seem to have a slight bluish haze, even when the engine is in good condition.

When troubleshooting a diesel, always note the colour of any exhaust smoke, as this is a powerful diagnostic tool.

Black smoke

Black smoke is caused by *all* the fuel being incompletely burnt due to insufficient air for complete combustion. This black smoke comprises small carbon particles, which, in extreme cases, can form an 'oil-like' layer on the surface of the water. Don't be misled, this is carbon and not oil.

As we discussed earlier, a diesel engine sucks in a full cylinder of air on each induction stroke. Fuel is added to produce the required power to overcome the load to achieve the selected rpm. If too much fuel is added, such that there is insufficient air in the cylinder to burn it, black smoke will result. This black smoke represents fuel injected but not producing any power, and so is fuel wasted. This can happen under several circumstances.

Acceleration black smoke

If the speed lever is opened too rapidly, full acceleration fuel will be injected. It takes time for the engine rpm to increase, and so initially there will be insufficient air to burn all this fuel. Black smoke will result. Open the speed lever slowly. Electronic fuel management prevents this by matching fuel injected to engine

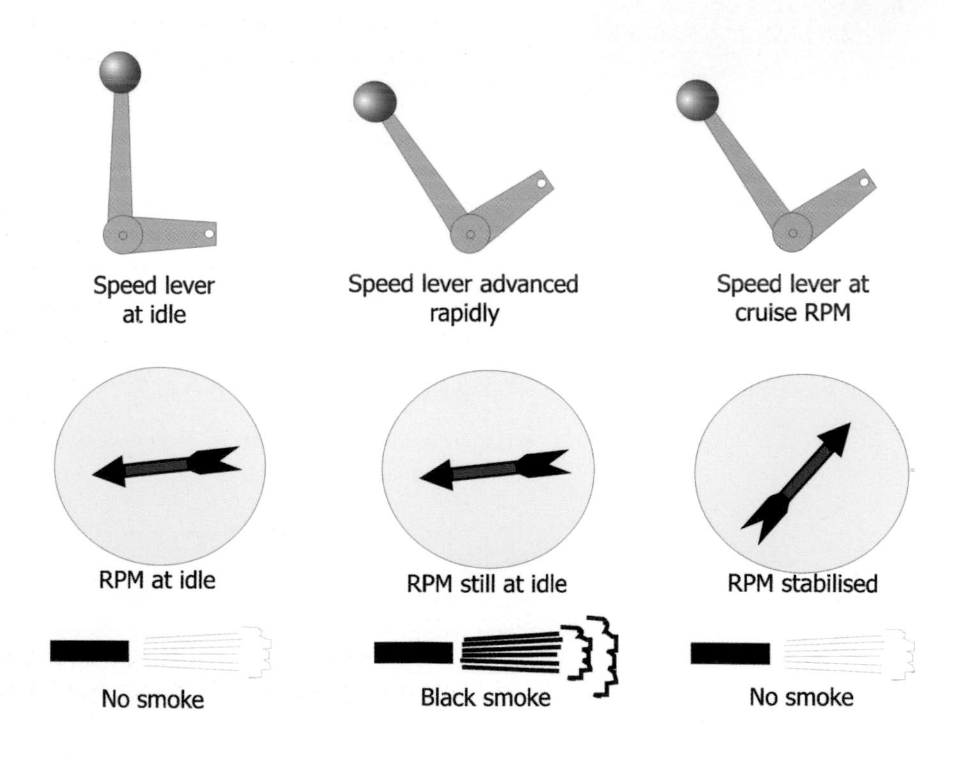

acceleration, resulting in a clean exhaust, however fast the speed lever is opened.

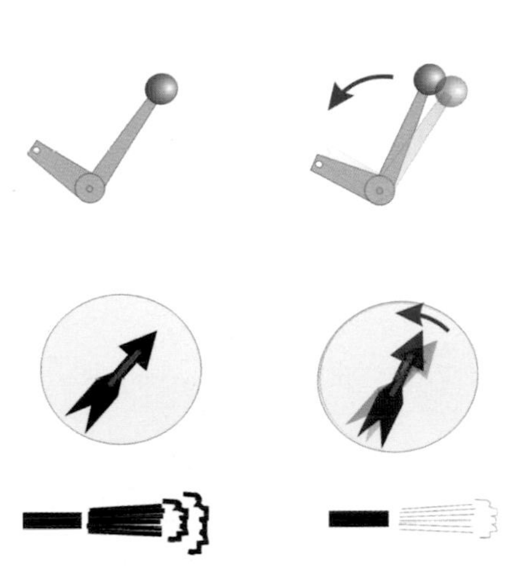

If overload black smoke occurs, ease back speed lever to reduce requested rpm and thus fuel input until black smoke ceases.

Overload black smoke

If more rpm is demanded than the engine has power to produce, black smoke will be produced. If the requested rpm is not achieved, acceleration fuel will be injected under the control of the governor, because the governor will not be 'in balance', even though the rpm is not accelerating. Because of the restricted rpm, insufficient air will be sucked in to provide complete combustion, resulting in black smoke. Closing the speed lever until the rpm *just* starts to reduce will ensure that there is sufficient air to burn the fuel and black smoke will cease, ensuring all the fuel injected is producing useful power, thus removing the overload condition.

Overload can result from a fouled hull or propeller, towing another boat, running in gear when moored or proceeding into heavy seas. With a marine engine it is due mainly to the use of a fixed pitch propeller. It is equivalent to driving up a hill in too high a gear in a car. Overloading can cause engine damage and in any case wastes fuel.

Other black smoke

Black smoke can be the result of insufficient air reaching the cylinders due to a restriction in the air supply. This restriction may be caused by a blocked air filter, or insufficient air getting into the engine compartment. This latter can be checked by opening the engine compartment and checking if the black smoke ceases.

On a turbo-charged engine, a faulty turbo-charger delivering insufficient air for combustion can cause black smoke.

One case I came across was caused by delamination of a rubber pipe connecting the air filter to the inlet manifold.

Blue smoke

Blue smoke comes in a number of different shades and hues. It ranges from a light hazy almost white colour through to dark blue. Watching a big diesel start shows a mixture of all possible hues!

Blue smoke is often an indication of burning lubrication oil in a worn engine, but fuel can give blue smoke as well. Blue smoke results because some of the fuel droplets (or lubricating oil) do not burn at all. This is because if a fuel droplet is too large in diameter it will not burn (this is a function of its surface area and mass). The size of the unburned droplet determines the shade of blue.

The fuel injected by a serviceable injector has droplets of the correct size to burn properly. When the engine is cold, some of this fuel may condense on cold surfaces of the combustion chamber and cylinder and form larger droplets that will not then burn. As the engine warms up, these condensed droplets will become smaller and, at some stage, will become small enough to burn. An engine running at light load may have some fuel condensation producing some droplets too large to burn, giving visible smoke.

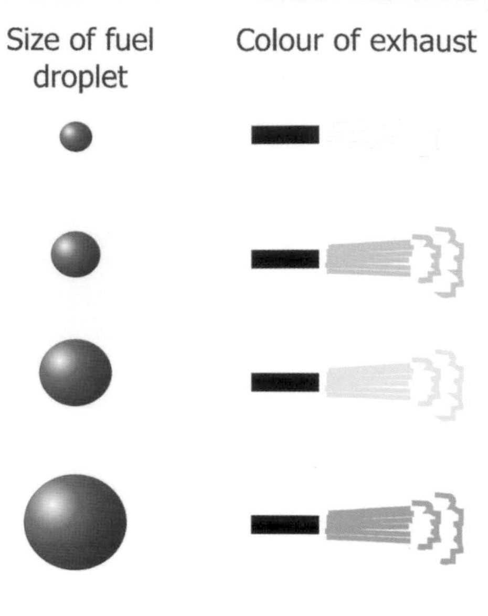

Effect of size of fuel droplet on colour of exhaust smoke

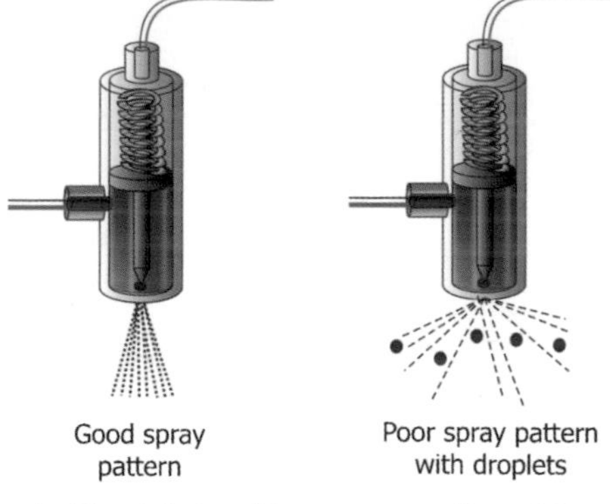

Worn injector with poor spray pattern giving rise to a smoky exhaust, poor starting and high fuel consumption

A worn injector may produce a deformed spray, producing some fuel droplets too large to burn.

A fuel injection pump with incorrect timing may supply some of the fuel to the injector too early or too late. Fuel supplied at the wrong time may reach the cylinder when the temperature is too low for combustion (too early) or there is too little oxygen left (too late). This fuel will be unburned and will result in blue smoke.

White smoke

White smoke is water vapour. Lots of white smoke is lots of water vapour!

When a car is first started on a cold, damp day you see white vapour coming from the exhaust. This is water vapour condensing in the exhaust, because for each gallon of petrol or diesel fuel burned, you get a gallon of water as a product of combustion.

On a boat with a water-cooled exhaust, the cooling water being injected into the exhaust system masks this relatively small amount of water vapour. Some people expect a cylinder-head gasket leak to show up as water vapour in the exhaust, but the quantity of water would be just too small to be seen on a boat.

Normally, a small engine would not have visible water vapour in the exhaust. A powerful engine being run at high speed, such as on a planing motor cruiser, would have a fair amount of water vapour visible in the exhaust, and this is quite normal.

Some books describe the smoke from fuel injection problems as being 'white'. To me this is actually 'light grey', but the difference is small. Watch a diesel, especially a big one, being started from cold and you will see quite a lot of varying colour smoke. The lightest of this is 'light grey'. You would not expect 'white' water vapour when starting from cold. Noting this should help to distinguish between the two.

CYLINDER BORE GLAZING

Certain operating conditions cause *cylinder bore glazing*, which reduces compression and increases oil consumption. Many yachtsmen deny the existence of this phenomenon, so I went to three of the world's leading engine manufacturers, who all came up with very similar descriptions of what happens and why.

Basically, they all said: 'don't run a diesel engine under light load – ideally aim for a minimum of 60% load – and don't run the engine too cool.' They all made the comment that working a diesel engine lightly and letting it warm up too slowly was actually being cruel, rather than kind, to the engine.

Reproduced with permission from Volvo.

Compression

The surface finish of the inside of the diesel engine cylinder is not highly polished, as one might imagine. It has a network of shallow grooves machined on the surface, reminiscent of a surface that has been lightly sanded in a regular manner. This is called *honing*.

The purpose of this honing is to hold lubricating oil to enhance the sealing of the piston rings to the cylinder wall, to stop air blowing past the piston rings during the compression stroke. Without the oil-filled honing grooves, the compression would be insufficient to achieve ignition temperature in a diesel engine.

Cylinder bore glazing occurs when these grooves become filled with carbonised lubricating oil, so that they are no longer able to hold liquid oil to enhance compression.

Initially the outer surface of these honed grooves is not completely uniform and forms a slightly 'rough' surface. This surface needs to be 'run in' during the first

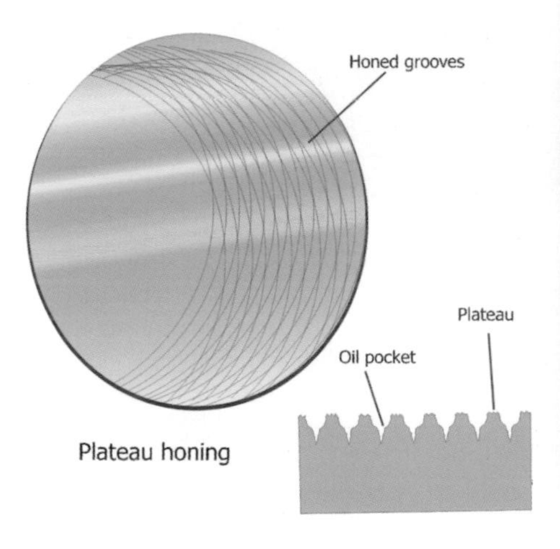

Plateau honing

Rough plateau

Run-in flat plateau

few hours of the engine's service to give a flat plateau between the grooves.

High pressure in combustion chamber gets behind compression ring and forces it aginst cylinder wall

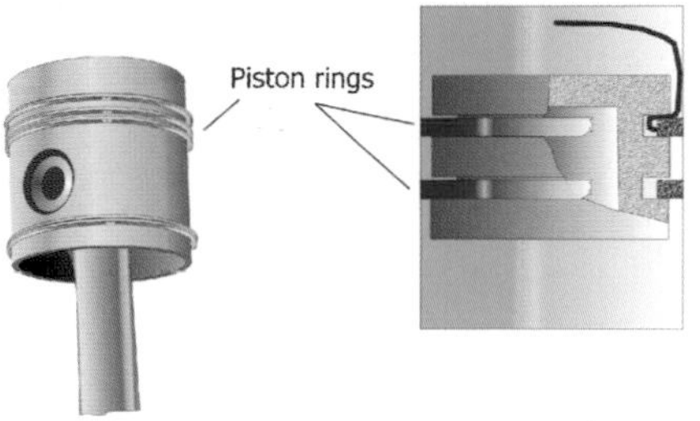

Glazing

The top, 'compression', piston ring is designed to allow combustion pressure to get behind it to push the ring outwards into firm contact with the cylinder wall.

If this pressure is insufficient, the corners of the honed grooves will become rounded instead of square, increasing oil consumption because oil will be able to migrate upwards towards the combustion chamber. As far as oil consumption is concerned, the first 40 hours of engine running are the most critical. New engines should be filled with running-in oil, which is replaced by normal lubricating oil, as specified by the engine manufacturer, at the first service. This running-in oil allows the honing to bed in properly.

Plateau with rounded corners due to running under low load

When dispatched from the factory to the boat builder, the engine may have no oil in the sump. The builder may then fill the engine with normal engine oil that will prolong the running-in period and in normal leisure use, the engine may never become properly run-in. Use of a higher grade of oil than specified for normal use may also prevent the engine from becoming properly run-in.

If this pressure is insufficient, combustion gases can pass between the cylinder wall and the piston ring, carbonising the oil in the honed grooves. This is an ongoing process throughout the life of the engine. Once the honing grooves are full of carbonised lubricating oil, compression is reduced, giving difficult starting.

Bore polishing

Prolonged cold running of the engine can result in considerable carbon build-up on the piston crown and the portion of the piston above the compression ring. This carbon can become baked onto the piston with hotter running, and it then forms an abrasive surface which wears the cylinder.

Prevention of bore glazing and polishing

Diesel engines should always be run under load so that combustion pressure is sufficient to force the piston rings into firm contact with the cylinder bores. The combustion pressure depends on the power being produced, which in turn depends on the fuel flow. Low fuel flow equals low combustion pressure.

- The propeller should be 'moving water'.
- Ideally, a marine diesel with a fixed pitch propeller should be cruised at around 70% (or more) of its rated rpm.
- 75% rpm with a fixed pitch propeller equates to about 50% engine power (or load).
- Motor-sailing at low rpm is bad – the engine should be contributing a reasonable amount of 'drive'.
- Battery charging in neutral at any rpm is bad – get the prop 'moving water', because even a high output (120 amp at 12 volt) alternator requires only around 3 hp to drive it.
- High rpm in neutral has low fuel flow and hence low combustion pressure – the engine is not under load.
- 'Warming up' at idle for more than two minutes is bad – if you need to warm up for longer, do it in gear with the prop moving water at around 1800 rpm if the local conditions allow.
- Don't use super oils in a leisure marine application unless the engine manufacturer recommends it.
- I once stated in *Practical Boat Owner* (henceforth PBO) that synthetic oil should not be used on a particular engine. I was at once challenged by the PR department of a major supplier of synthetic oil, who said that their product was the proper oil to use. I asked him to confirm that they would indemnify the user against damage when used in the marine leisure environment and all went very quiet!

Whether or not bore glazing will occur in a particular engine seems to some extent to be influenced by other factors as well. I have known engines to suffer after all the preventative rules have been followed and also the reverse, where terribly abused engines seem to come to no harm. However, following the rules will give you the best chance of avoiding bore glazing.

Should your engine's bores become glazed, they can be deglazed by re-honing. This requires engine removal and removal of the pistons from the cylinders so that the proper procedure may be applied. In the early stages of the problem, changing to running-in oil for the next 50 hours and running the engine as hard as possible may arrest the progress of the problem, but again, it may not.

Re-honing

This is best left to the expert, who will use a re-honing tool. This is used in a powerful, slow-revving hand-drill. Using a suitable 'honing fluid', such as brake fluid, the tool is used in the bore in a reciprocating motion to cut new honing grooves.

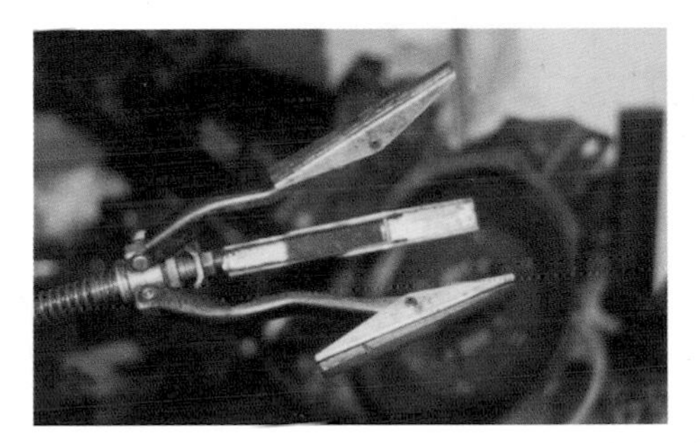

ENGINE POWER

There is a great difference between the power an engine is capable of delivering and what it is actually delivering at any time.

The marine engine manufacturer gives a graph of engine power and rpm. On this there will be two curves:

one showing what the engine is capable of delivering at full load at all rpm, and a second showing what a fixed pitch propeller can harness if it is matched correctly to the boat's maximum speed and engine's maximum power and rpm. This curve is known technically as the *propeller law power*. Different authorities use different shaped curves, as it's an inexact science and the curve will, in any case, be specific to the design of propeller. We will see that at 75% rpm, the propeller is harnessing only about 50% of the engine's maximum power. When cruising, you won't be far wrong if you cruise at the rpm giving maximum torque.

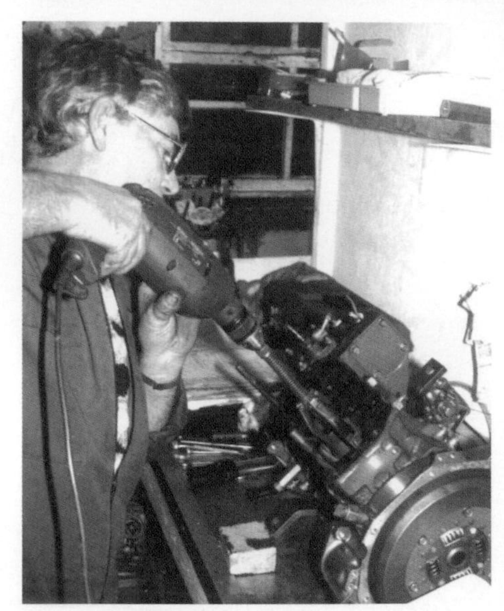

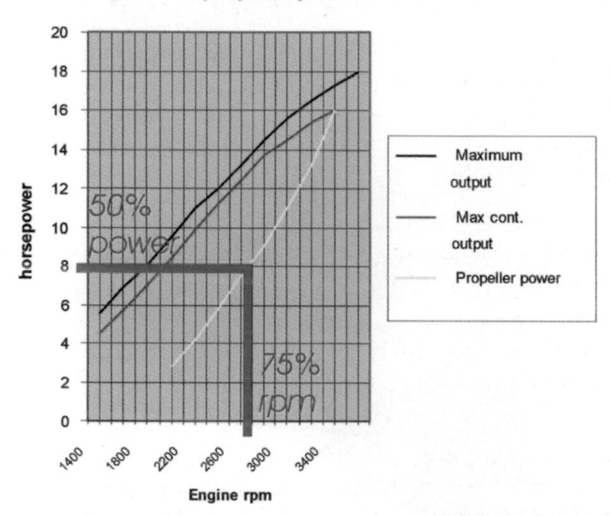

Some manufacturers give other graphs as well. *The specific fuel consumption curve* is a means of estimating the fuel consumption at different rpm. It gives fuel consumption per hour and doesn't take into account boat speed, so can't tell you how far you can travel. It's only fully valid when the engine is producing its maximum achievable power at any given rpm, but as this is the only information ever available, you have to use this when applying the propeller law power. *The torque curve* indicates the 'force' turning the propeller. A 'flat' curve indicates that the engine is more capable of accelerating the engine from low speed when under load.

What the manufacturer's figures mean

These performance curves are taken from the Yanmar 2GM20F data sheet

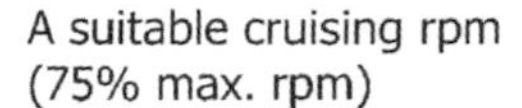

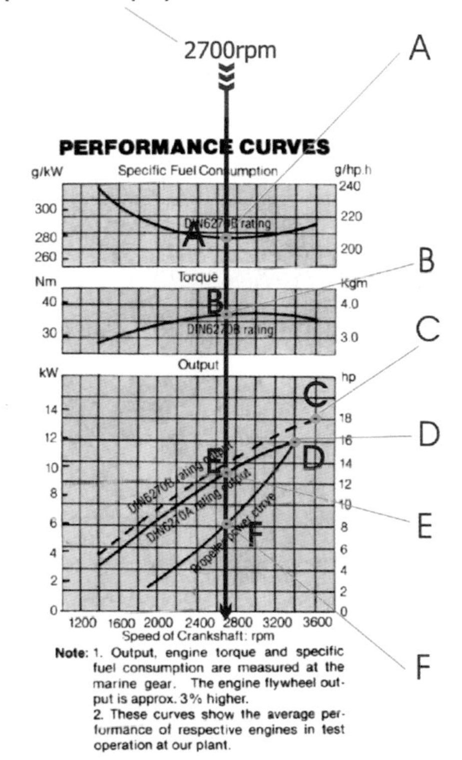

A SPECIFIC FUEL CONSUMPTION.
To get real fuel consumption you need to multiply this by the horsepower being absorbed by the prop.
In this case, looking at the bottom curve (F) we see that at 2700 rpm, the prop power is 8 hp. So, multiply 208 grams per hour by 8 hp and you get 1664 grams per hour. This works out as 0.46 gallons per hour.

B TORQUE
Don't worry too much about this.

C RATED POWER
The maximum power of the engine and the rpm that it's achieved. 18 hp at 3600 rpm

D MAXIMUM CONTINUOUS POWER
The maximum power (and speed) you can use continuously. 16 hp at 3400 rpm

E POWER CURVE
This is the maximum continuous power the engine can develop at any given engine speed.
Here it's 13 hp at 2700 rpm.

F PROPELLER CURVE
Because a fixed pitch prop is fully efficient only at its designed speed (in this case 3400 rpm), it can't absorb all the engine's output at any other speed.
Here it can harness only 8 hp at 2700 rpm (and not the full 13 hp the engine is capable of at 2700 rpm), so only 8 hp is available to drive the boat. Fortunately, it's using only 8 hp's worth of fuel as well.

NOTE:

Yanmar gives output at the gearbox, most others give output at the crankshaft and overstate their power by 3%. A further 3 to 5% needs to be deducted to allow for the stern gear.

Manufacturer's performance curves for YANMAR 4LH-DTE 160 hp diesel

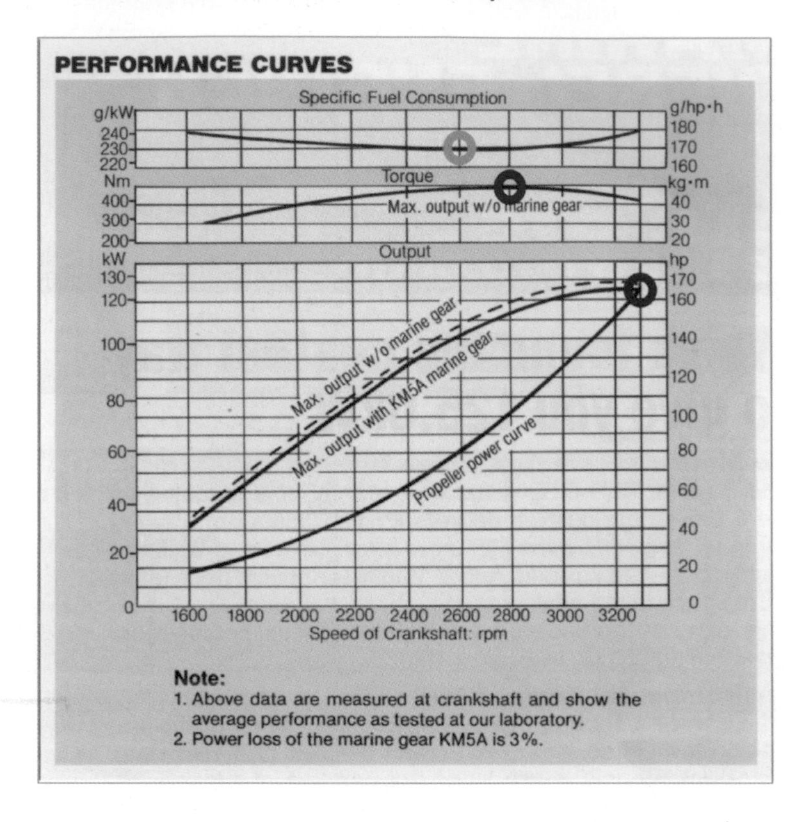

- RPM for best specific fuel consumption
- RPM for best torque
- Maximum power

Motor cruisers

The engine handbook tells you the maximum engine speed. This is normally for a maximum of one hour. For longer times than this you must observe the 'maximum continuous' rpm. In the Yanmar engine curve shown, this is 3200 rpm, only 100 rpm less than maximum.

However, for maximum cruising speed it is normal to 'throttle back' by at least 300 rpm from the maximum that can be achieved on a power boat. This will equate to approximately 75% power. Endeavour not to cruise at less than 75% maximum rpm.

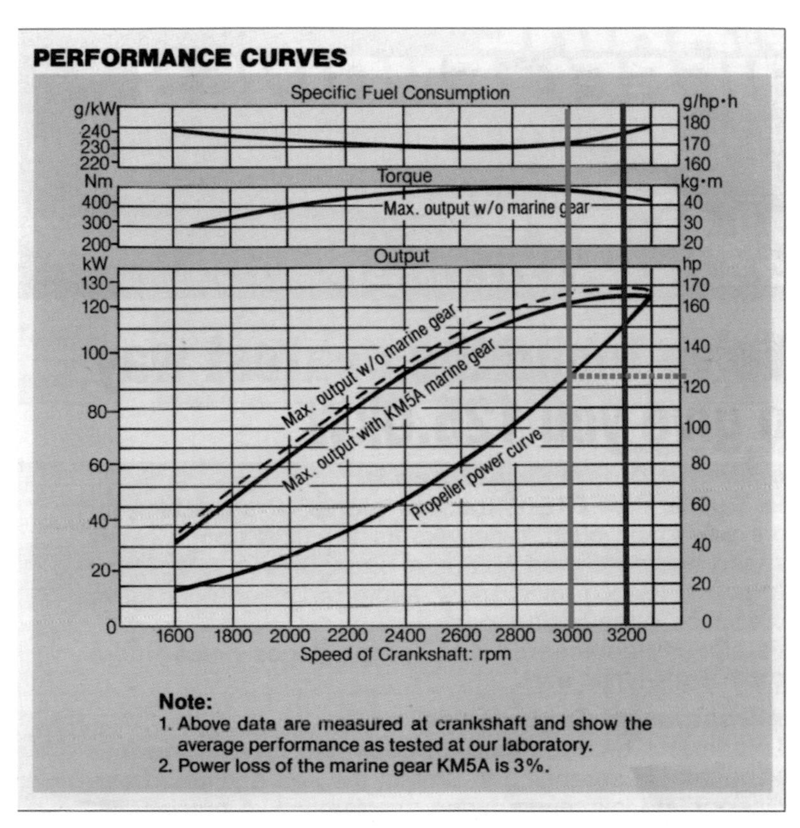

Normal and maximum cruising-power

Sailing yachts

It should be your aim to cruise at around 75% maximum rpm, i.e. about 50% power.

Modern sailing boats are often fitted with a more powerful engine than is desirable.

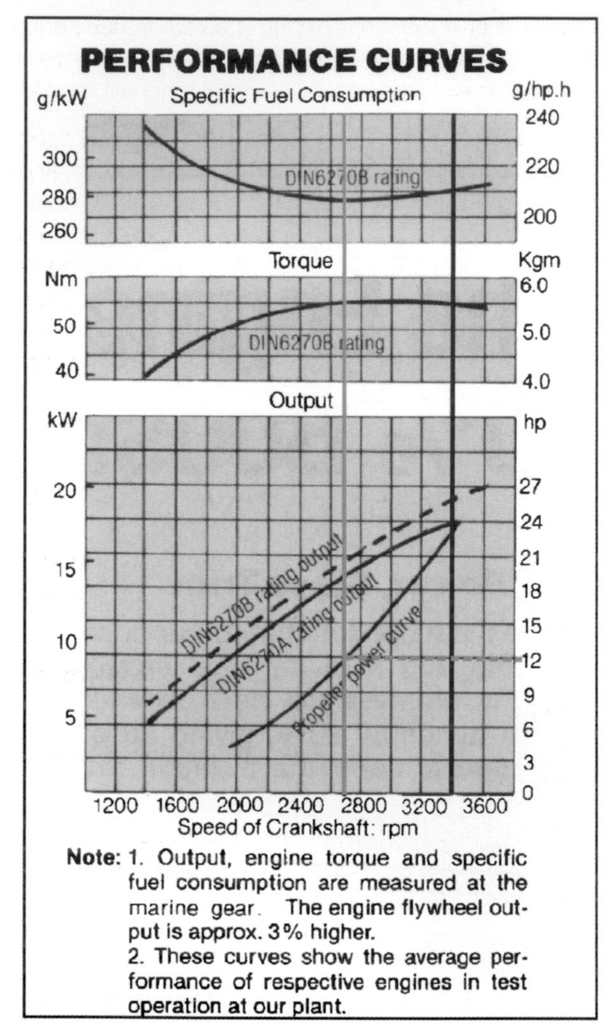

Cruising and maximum power – sail

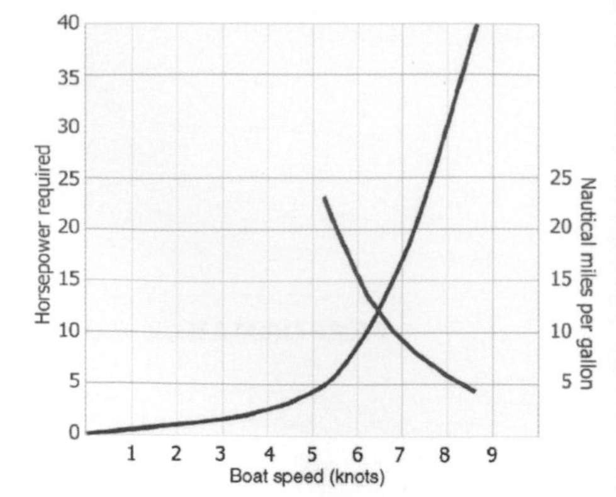

Power required to propel an easily driven 36-foot sailing boat of 6 tonnes displacement in calm conditions. Cruising at 6.3 knots needs 10 horsepower, so at half power for cruise you need a 20 horsepower engine installed. This gives a maximum speed of 7.2 knots in calm conditions. The cruising fuel consumption would be around 12 mpg. Fit a 30 hp engine and you have to cruise at 6.8 knots for half power and get only 10 mpg, or cruise at only 33% power. Go up to 40 hp and things just get silly, with 7.2 knots and 8 mpg, or cruising at only 25% power.

FUEL CONSUMPTION

Fuel consumption can be deduced from the engine maker's curves, which will give the *specific fuel consumption* (SFC). Multiply the SFC (often in grams per hour) by the *propeller horsepower* to get the consumption per hour at that rpm. If you've got two engines, multiply that by two. A gallon of diesel weighs about 8 lbs – about 4.2 kg (4200 g) and there are about 4.5 litres to a gallon, so you can do the sums in your units of choice.

There's a very convenient rule of thumb, so you don't need to get involved with the graphs. A diesel engine uses around 5 gallons per hour for every 100 horsepower. If you cruise at 75% rpm, that's about 50% maximum power. So, for every 100 rated (maximum) horsepower, you should use two and a half gallons per hour at cruising rpm. If you average much less than that, you are not loading your engine really sufficiently.

OPERATION OF THE ENGINE

You can prolong the engine's life and ensure maximum reliability by operating it correctly.

- Warm the engine up for only a couple of minutes if you are initially going to use only low power – you need 60 °C for full power.
- Once you have started the engine, don't stop it until the engine is warm (about 10 minutes or so).
- Run the engine under load – ideally cruise at around 75% rpm on a displacement boat.
- Planing boats will need to be cruised at around 300 rpm less than maximum to obtain optimum performance.
- Open the 'throttle' slowly.
- When changing gear, pause in neutral (count 1 second when in neutral).
- On a twin engine boat, if forced to run at slow speed, run on one engine if possible.
- Run the engine at idle for a couple of minutes prior to shutting it down, but don't leave it running while you tidy up the boat.

READ THE BOOK

Men tend not to read instruction books, yet your diesel engine handbook contains essential information and will not take long to read. I have come across a huge number of Volvo Penta 2000 series engine owners who are completely unaware of the correct cold start procedure, due entirely to their unwillingness to read the book.

A few questions may well illustrate the point:

- What is the cold start procedure for your engine?
- When sailing should you lock the prop shaft?
- How long should you remain in neutral when changing from forward to reverse?
- What is the normal oil pressure?
- What is the normal water temperature?
- Do you have any engine anodes; if so, where are they?
- What engine oil and gearbox oil should be used – are they different?
- Where should you check the cooling water level?
- Should the engine alarm sound when you stop the engine?

WHERE IT'S ALL GOING

We saw at the start of the book how simple the concept of the diesel engine is. Indeed, for many years the diesel maintained this simplicity, so that maintenance, troubleshooting and repair were comparatively easy.

Today, marine diesels of up to about 50 hp retain this simplicity, but from then up, complexity is added for the sake of greater efficiency, greater power from small packages and better emission control, so that today's larger engines are difficult and expensive to maintain, difficult to troubleshoot and expensive to repair.

For the impecunious or DIY yachtsman, or those straying from places where repair facilities abound, this decline in simplicity will become more burdensome.

The Fuel System

The fuel system comprises everything from the boat's fuel tank to the engine's fuel injector(s).

There are two distinct parts of the fuel system: that installed by the boat builder and that attached to the engine. The tank, primary filter and all the pipe work to and from the engine are designed and installed by the boat builder. Sometimes they don't make a very good job. On older boats, you may find modifications to this part of the system, so investigate it thoroughly to see if you need to bring it up to scratch. Generally, apart from routine servicing and fair wear and tear, there should be little trouble from anything supplied with the engine.

THE FUEL TANK

Modern fuel tanks tend to have all their connections at the top of the tank. There seem to be two reasons for this. The first is the mistaken belief that this is required by the various regulating bodies – this really applies only to inland waterways; the second is that it makes life easier for the boat builder. Ideally, the fuel tank should have a sludge/dirt trap at the lowest part of the tank. This will have its own drain point in the

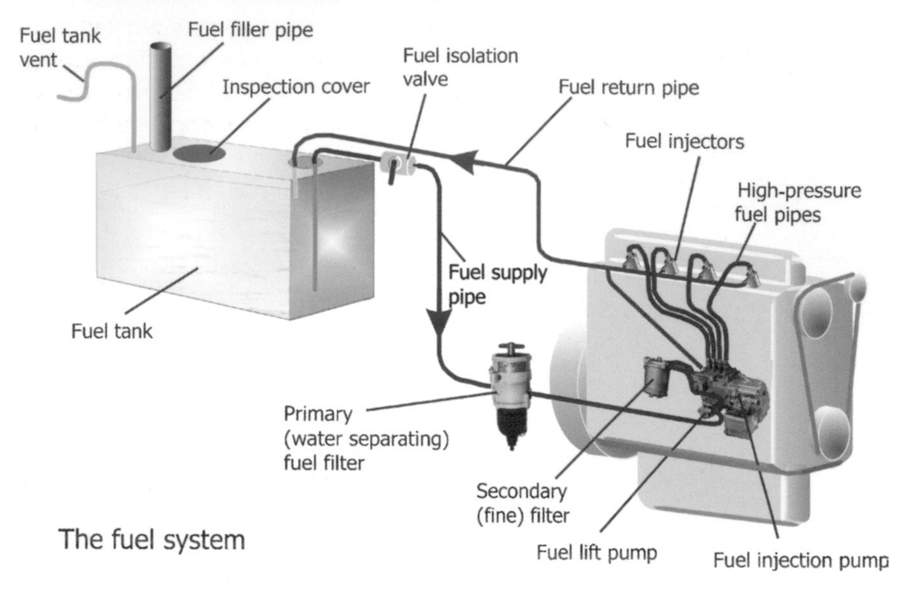

The fuel system

event that dirt or water accumulates at the bottom of the tank.

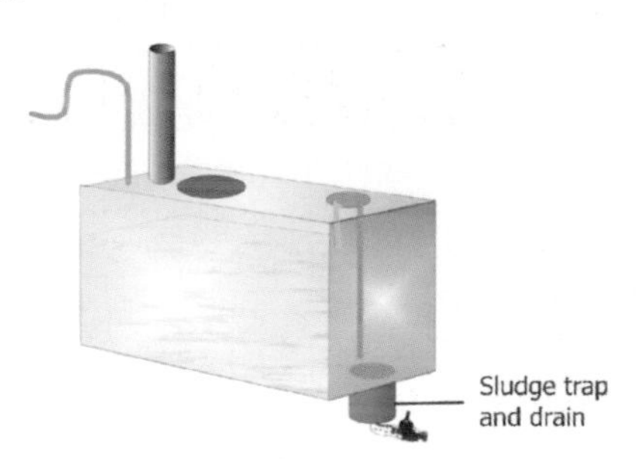

Ideal fuel tank

Ideally, the drain point should have a drain cock, but the cock outlet must have a blanking plug.

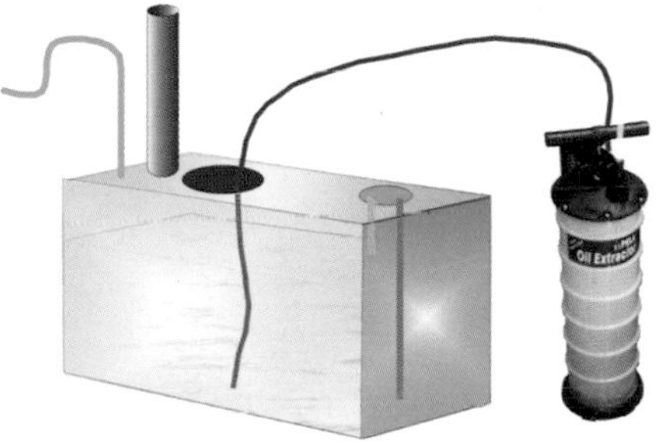

Empty tank through inspection hatch

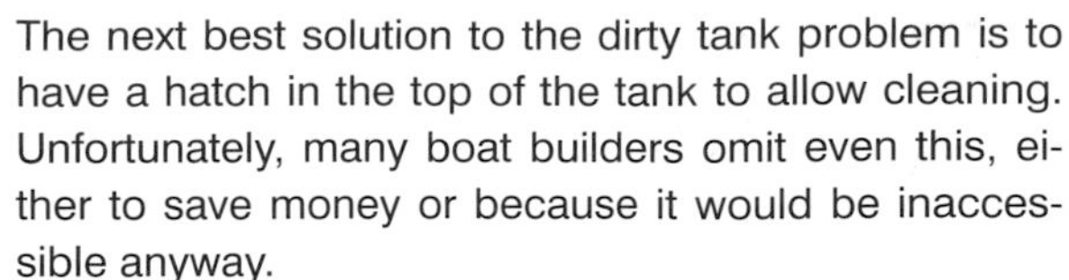

The next best solution to the dirty tank problem is to have a hatch in the top of the tank to allow cleaning. Unfortunately, many boat builders omit even this, either to save money or because it would be inaccessible anyway.

It is generally recommended that fuel tanks should be cleaned every five years. On most boats it is difficult, if not impossible, to remove the fuel tank, unlike on the illustrated Westerly, which, as you can see, also has a sludge trap.

The fuel tank may be made of mild steel, stainless steel or polypropylene.

- Mild steel is cheap and easy to weld, but suffers from internal and external corrosion.
- Stainless steel doesn't suffer from corrosion, but the welds can suffer from crevice corrosion and the welds are the weak point of this material. They are more expensive.
- Polypropylene is robust, is translucent, so you can see the contents, and it doesn't corrode. Cost is much the same as stainless steel. It is my material of choice.

The outline of the tank baffles can be seen on this tank

Tanks bigger than about 20 gallons (90 litres) need to have baffles fitted to reduce the 'sloshing around' of the fuel inside. The position of the baffles can be seen clearly on the illustrated mild steel tank ready for repainting.

Draining the tank using the engine supply pipe will not empty the tank as the supply pipe doesn't reach the bottom of the tank. This is to prevent the supply pipe picking up dirt from the bottom, but of course this is effective only in still water. At sea, any dirt will be shaken up.

It's been reported that some tank makers have fitted a strainer to the supply pipe, inside the tank. Owners of such tanks will be unaware of its existence and may be unable to clean it should it become blocked.

Fuel tank isolation valve

It must be possible to isolate the fuel tank (turn off the fuel). This mechanism should be fitted as close to the tank as possible and should be accessible from outside the engine compartment. I don't like gate valves, as you can't tell if they are on or off. Use a lever valve.

Gate valve

Lever valve

Tip

If you have no easy way of cleaning the inside of your tank, you will need to improvise. By removing the filler pipe from the top of the tank, you may be able to pass a suction tube inside the tank to suck out some of the muck. This is more likely to be effective in removing water than dirt.

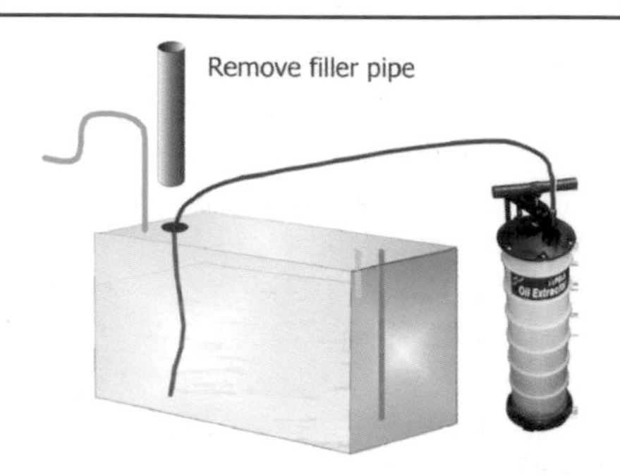

Empty tank through tank filler

Tips

- Some boats, such as newer Moodys, may have a solenoid cut-off valve fitted immediately after the manual valve. This valve may be operated by the 'ignition switch', so you will get no fuel flowing unless the 'ignition' is switched on.
- I once had a boat with the valve inside the engine compartment. I fitted a replacement choke cable from a car accessory shop so that the valve could be turned off from the cockpit locker.
- A boat I worked on had no fuel valve at all. It did, however, have a short length of rubber pipe from the tank fitting to the solid pipe, so I was able

temporarily to shut off the fuel using a woodworking cramp. What it needed was a valve fitted in place of the rubber hose.

- Electrical fuel content gauges are notoriously unreliable. A transparent sight gauge mounted on the side of the fuel tank is simple and fool-proof. However, surveyors, quite rightly, disapprove of them as they can fail and then you lose all your fuel into the bilge. Central heating oil tanks have a push-to-open valve at the bottom of their sight tubes and you can incorporate one of these easily into your system. If you can't get one locally, try www.asap-supplies.com It's unlikely that you will be allowed to use this on a UK inland waterways boat.

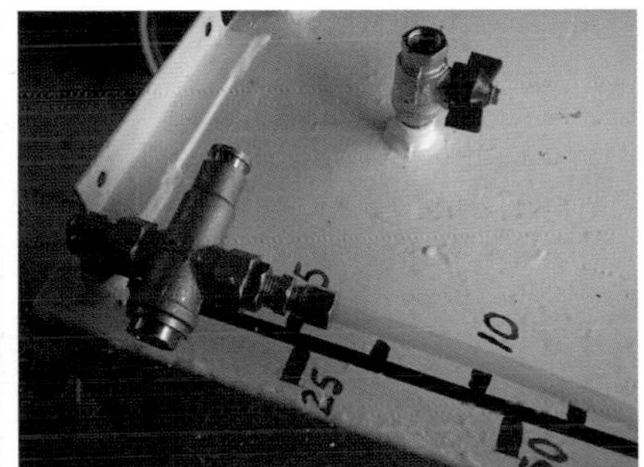

Push valve

- On some installations the boat builder has fitted a mesh filter to the fuel outlet on the inside of the tank. Stupidly, there's no way of getting to it and there's no indication that it's there. If you find that there's an inexplicable blockage at the tank exit, this might be the problem. The solution will depend on the installation, but compressed air may help, or you may have to cut an access panel into the top of the tank.

PRIMARY FILTER

This filter goes under several different names, such as *primary filter*, *pre-filter*, *water separating pre-filter* or *agglomerator*.

Why it's fitted

The filter has two jobs: one is to keep any fuel tank dirt from passing further down the fuel system, and the other is to collect any water which may be in the tank to stop it reaching the engine. Ideally, it should have a transparent bowl, so that water and dirt can be seen without having to drain the bowl. Some people believe that you are not allowed a transparent bowl, but this isn't true. Diesel fuel systems are always allowed a transparent bowl, even on inland waterways – where it must be fireproof.

Where it's fitted

It's best fitted as close to the tank isolating valve as possible to prevent blockage further along the line. The overriding consideration must be accessibility.

Draining water

1. There's no need to turn off the fuel.
2. Put a container under the filter to catch the water/fuel drained off.

3. Unscrew the drain valve at the bottom – this won't allow air into the system.

4. Close the valve when the water is removed.

Servicing it

You will need to remove the bowl to clean and replace the filter. Lorries have a vacuum gauge to indicate when it needs to be serviced, but you'll rarely find this on a small boat. I suggest that you use the same schedule as for the engine filter.

1. Turn off the fuel at the tank.

2. Drain the filter using the water drain plug at the bottom – you may have to loosen the bleed screw to do this.

3. If there's a central bolt through the whole filter assembly, remove this bolt, holding on to the bottom of the assembly as you do so.

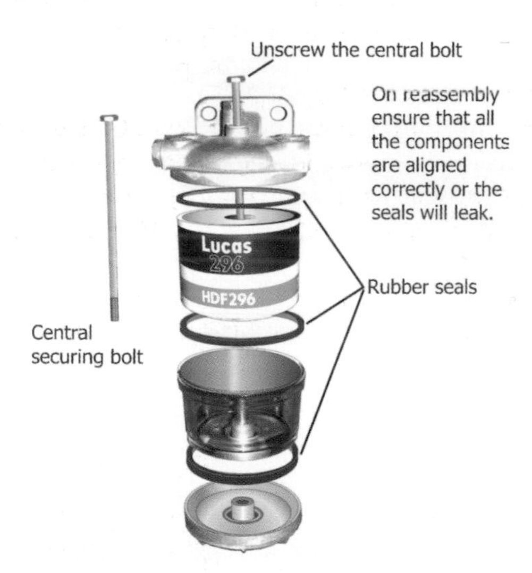

4. If the bowl and the filter unscrew from the filter body, unscrew them from the body.

5. If it's a 'T' handle filter, unscrew the handle and remove the filter.

Tips

- If you don't have a transparent bowl, examine the container to check if water is present and close the valve when fuel starts to drain.

- Get a transparent bowl for your filter, as you'll need to check this daily if you can't see if water is present!

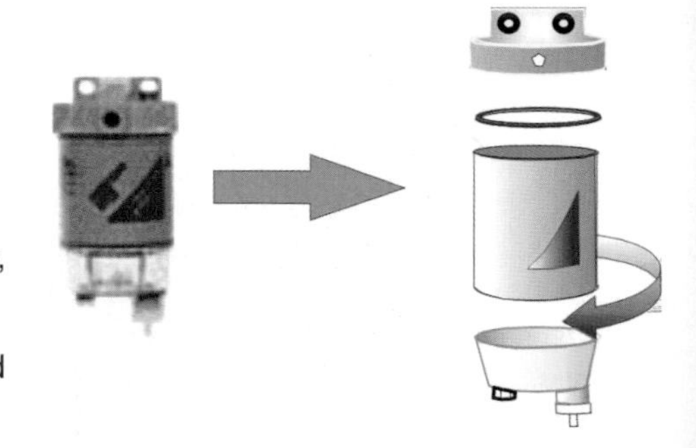

Tips

- Lucas and CAV type filters with a single central bolt need particular care when you reassemble them, as unless they're assembled absolutely square, they will leak when you refill them with fuel.
- Many older boats were not fitted with one of these, so my advice is: if you haven't got one, fit one. It should be fitted in an accessible place so that it can be checked easily, as it's on the daily checklist.

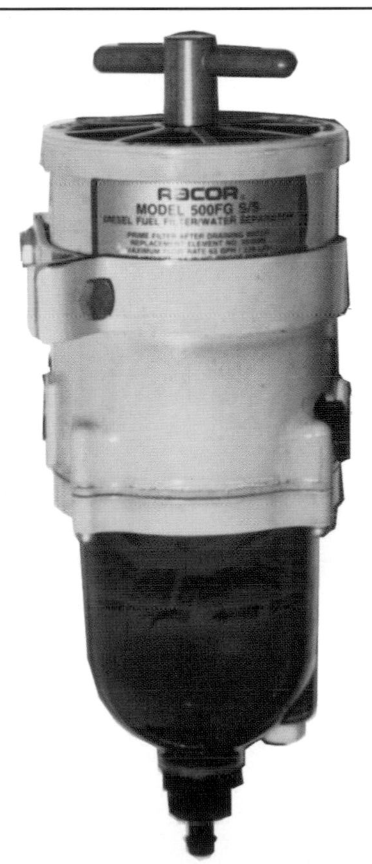

6. Clean the filter bowl – if it's dirty, this indicates that there's dirt in the tank.
7. Replace the filter with a new one.
   — Lubricate any *flat* rubber sealing washers with fuel before reassembly.
   — The new filter should be supplied with all the new seals required. Make sure you use them.

FUEL LIFT PUMP

Unless the fuel system is totally gravity feed – unusual in a marine installation – there will be a low-pressure pump to transfer fuel from the tank to the engine. This is commonly called the *fuel lift pump*.

Where its fitted

On smaller engines this is usually a separate unit mounted on the side of the engine and is driven by the camshaft.

Larger engines usually have the lift pump mounted on, or incorporated within, the fuel injection pump.

Separate fuel lift pumps

Prior to about 1988, these pumps could be taken apart to have their valves and diaphragm replaced. They also incorporated a coarse, nylon filter element. Often, this was the only filter prior to the engine's fine filter.

Cleaning the pump filter

This can be done without removing the pump:

1. Unscrew the top retaining bolt.

2. Remove the top cover.

3. Remove the filter.

4. Clean the filter in clean diesel fuel.

5. Some pumps have a cylindrical filter.

Removing the lift pump

1. Unscrew the two attachment bolts – usually these are Allen screws, so you'll need a hexagonal wrench.

2. Remove the bolts.

Tip

If the filter cover hasn't been removed for some time, it may not come off easily.

- Put the screw back into the cover at an angle.

- Use the screw to lever the cover off.

3. Remove the pump.

4. Remove the gasket – in this case a rubber 'O' ring.

5. The pump removed.

Pumps vary in their design. The exploded drawing is of a typical type, and they usually have six small screws holding the diaphragm chamber together. Removing these screws allows access to its interior, so that the valves and diaphragm can be changed.

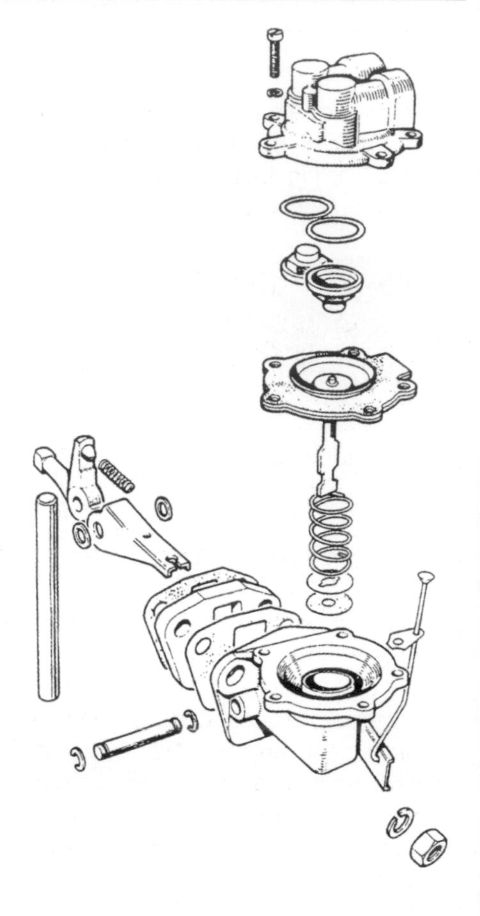

1. Remove the screws.

2. Pull the pump body apart to reveal the diaphragm. Note the diaphragm also forms the seal.

3. As pumps differ, you'll need to explore how the diaphragm is removed.
4. The valves are often held in place by a retaining plate attached by one or two screws.

Tip

Some valves are a push-fit in the housing and are then retained by a portion of the casting that has been burred over to hold it in place.

These are more difficult to remove and refit. If the valve has failed, then there's no harm if you damage it when you remove it, provided you have a replacement. Remove the burrs with a needle file, fit the new valve and use a centre-punch to form punch marks to ensure the valve stays in place. This may not be the best engineering practice, but it does work – I used this method on a Volvo Penta pump and suffered no recurring problems.

5. Unscrew the retaining plate screw(s).

6. Remove the valve(s). This is the inlet valve.

7. On refitting, ensure that the two valves are fitted the correct way round – one allows fuel in and the other fuel out.

Later type pumps

Later pumps cannot be serviced and have no filter. Usually, these newer pumps can directly replace the older type.

Fuel priming lever

Fuel pumps have a priming lever so that the pump can be operated without the engine running. Sometimes the pump is mounted in such a position that the priming lever is not easily accessible. In this case there may be an extension so that the pump can be primed.

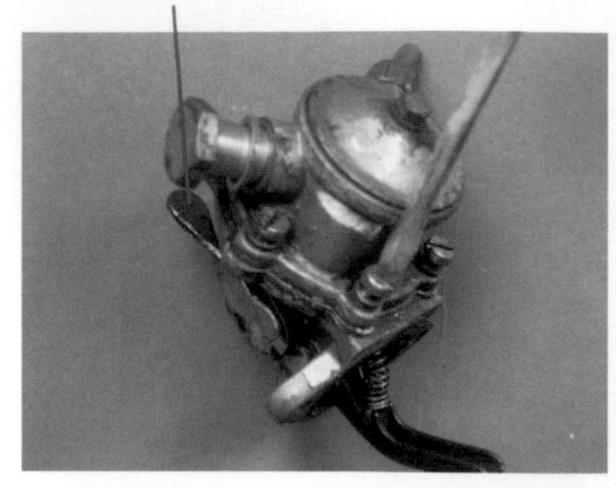

Pumps which are attached to the fuel injection pump

Sometimes the pump is of the type described previously and can be removed. Otherwise, any work is a job for an injection specialist.

ENGINE FINE FILTER

A fine filter is attached to the engine by the engine manufacturer. In times past it would have been the only filter, but now it's the second in line, after the pre-filter. As part of the engine, the quality (fineness) is specified by the engine manufacturer, as is its service schedule.

The filter may have a bowl, inside of which a replaceable filter element is fitted, or the filter may be of the 'spin-off' type.

They come in all shapes and sizes.

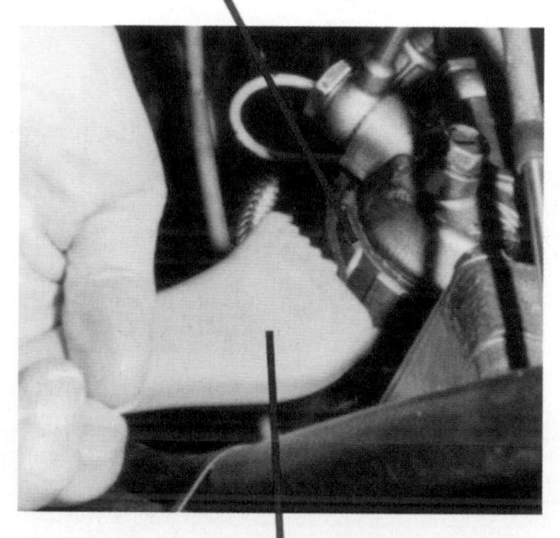

Tips

- The underside of the filter head will have an 'O' ring seal in a deep groove. Remove this by using a small screwdriver or similar. The replacement ring will seem too large a diameter and part will hang down when you insert it into the groove. Work your way round the seal so that you have a number of smaller 'hangy-downy' bits. Then work these into place one by one.
- Some bowl-type filters have a threaded retaining ring which you unscrew to release the filter, rather than a central bolt. You may need to use a 'strap wrench' to loosen the ring.

Servicing the fine filter – bowl type

1. Turn off the fuel isolation valve.
2. Unscrew the central bolt holding the filter bowl to the filter head.
3. Ensure that you hold the filter bowl so that it won't fall.
4. The bowl will be full of fuel, so have a container ready into which you can empty the contents.
5. Remove the filter element from the bowl.
6. The replacement element will have all the necessary replacement seals.
7. Reassemble the components.

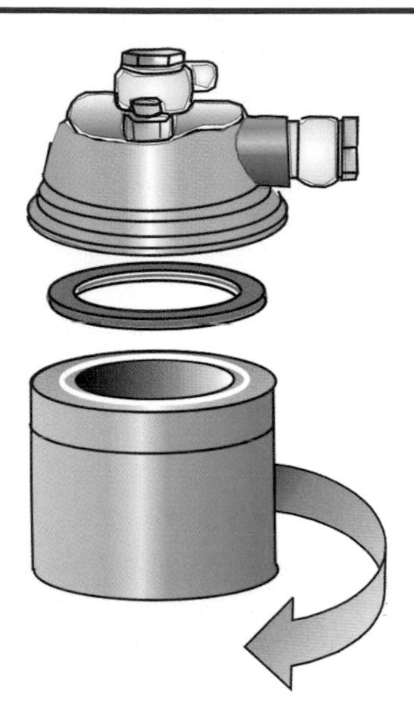

Servicing the fine filter – 'spin-off' type

1. Use a strap wrench to loosen the filter element.
2. Unscrew the element.
3. Drain the element into a container.
4. Lubricate the faces of the new sealing ring with fuel.
5. Screw the new filter into place.
6. Tighten the filter.

FUEL INJECTION PUMP

Servicing the fuel injection pump should be carried out by a fuel injection specialist. Although an expensive piece of equipment, it's very reliable and rarely needs attention, provided you feed it clean fuel without any water content. Only after having eliminated all other causes should you consider the injection pump as a cause of the symptoms.

- Light grey smoke is an indication of an injection problem – see the section on smoke in the previous chapter.
- Wear in the drive system of the injection pump could upset the injection timing.
- Injection pumps driven directly by the engine cam shaft have the timing adjusted by removing or inserting shims under the pump body.

- Externally mounted injection pumps can have their timing adjusted externally – these will have alignment marks.
- ANY adjustment of the injection timing must be carried out in conjunction with the engine workshop manual.

Where it's fitted

The fuel injection pump is often hidden by other equipment or on the side of the engine. Often, the easiest way to find it is to follow the high-pressure fuel pipes back from the injectors.

Fuel injection pump

Fuel injection pump

Reproduced with permission from Volvo

FUEL INJECTORS

High pressure (up to 200 atmospheres – 3000 psi) fuel is delivered to the injectors, which are sometimes known as atomisers, by steel pipes.

Injection pipe clamps
Fuel injectors
High-pressure pipes

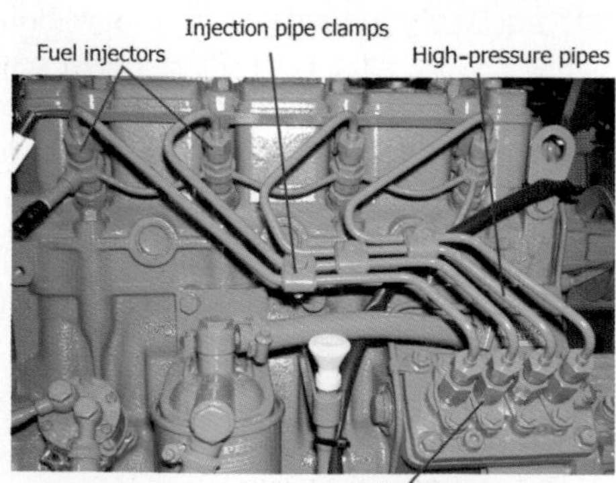

Fuel injection pump

These pipes are often of tuned length on high-speed diesels and any replacement must be of the same length.

The pipes are usually fitted with clamps. These are to damp vibration of the pipes and if not fitted, pipe failure is likely within 10 hours of engine running.

Most service engineers consider that injectors need to be serviced only when accompanied by symptoms of deterioration, as follows:

- continuous light grey smoke at normal running temperature and under load, indicating that some of the fuel is not being burnt – see the section on smoke; in the previous chapter;
- difficult starting from cold;
- sometimes accompanied by a rise in lubricating oil level due to unburned fuel dribbling down the cylinder walls into the sump.

Injectors are sometimes difficult to remove (a reason advanced by some as to why they should be removed annually). Some are unscrewed directly from the cylinder head, some are held in place by a clamp, and some engines have the injector mounted in a copper sleeve as a heat sink. This sleeve is in the cooling jacket and is in direct contact with the cooling water. Should it come loose while you are removing the injector, water

Screw-in injector
Screw-in sleeve
Heat-sink
Seal

will enter the cylinder and you will need to remove the cylinder head.

Engineers may have an impact tool which is placed under the head of the injector and hammered upwards to loosen the injector.

REMOVING AIR FROM THE FUEL SYSTEM

Air can enter the fuel system in various ways:

- a low level of fuel in the fuel tank allows air to enter the supply pipe in rough seas;
- air enters the system at a loose or leaking joint;
- in the course of servicing the fuel system.

The process of removing the air is known as *bleeding*, *priming* or *venting* the system. Filters and some injection pumps have *bleeding points* or *bleed screws*. During the course of servicing the fuel filters, it's usual to bleed them in turn after they have been changed, provided that the bleed points are above the level of fuel in the tank, i.e. they are gravity fed.

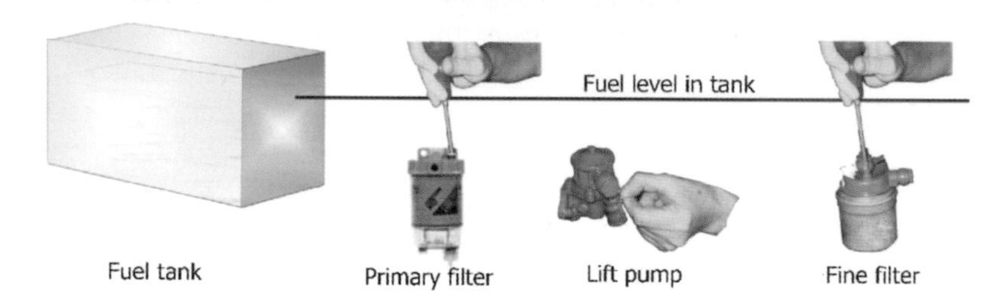

1. Open the fuel isolation valve.
2. Open the bleed screw on the primary (water separating) filter
   — first, only air will flow from the bleed screw,
   — then, a mixture of air and fuel in the form of bubbles,
   — lastly only fuel will flow.
3. Close the bleed screw.
4. Open the bleed screw on the engine fine filter.

5. Wait until only fuel flows from the bleed screw.
6. Close the bleed screw.

If the primary filter is below the level of the fuel in the tank and the fine filter above:

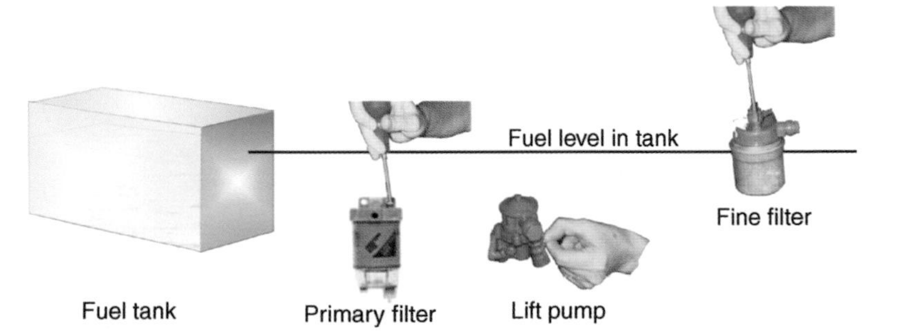

1. Bleed the primary filter as above.
2. Open the bleed screw on the fine filter.
3. Operate the fuel priming lever (plunger) until fuel flows from the filter's bleed point.
4. Close the bleed screw.

If both filters are above the level of the fuel in the tank:

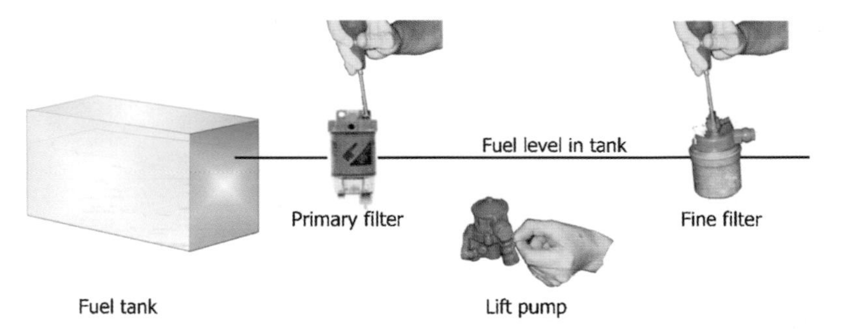

1. Don't attempt to bleed the primary filter.
2. Open the fine filter's bleed screw.
3. Use the pump's primer to expel the air from the whole system.
4. Close the bleed screw.

If there's a bleed point on the fuel injection pump – most modern pumps are self-bleeding:

1. Open the injection pump bleed screw.
2. Use the fuel lift pump's primer to remove air from the system between the fine filter and the injection pump.
3. Close the bleed screw.

If the engine stops or won't start and you suspect air in the system, bleed the fine filter.

STARTING AND STOPPING THE ENGINE

A diesel engine is stopped by cutting off the fuel at the fuel injection pump. This is achieved by movement of the fuel control rack to its *cut-off* position.

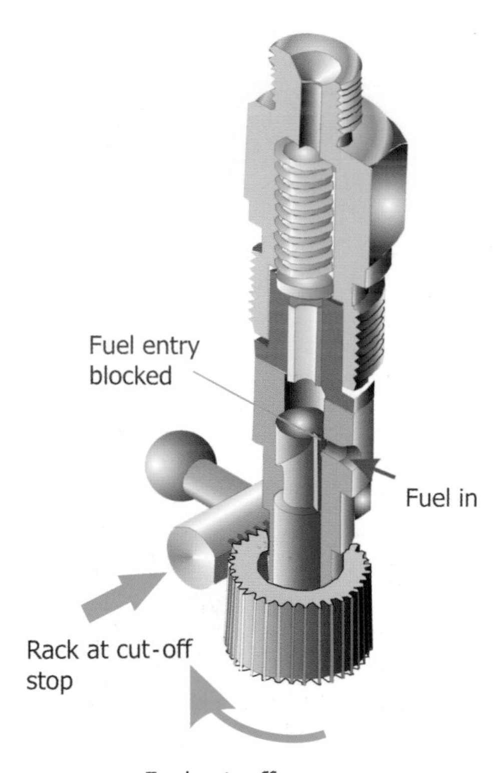

Tips

- Ensure that the liquid flowing from the bleed point is fuel and not water. When rubbed between the thumb and finger, fuel feels slippery, water doesn't.
- To ensure that you recognise your bleed points quickly, paint them a distinctive colour.

Painted bleed screw

- Keep the spanners or screwdrivers needed to bleed the system mounted securely in the engine bay so you know where to find them.

- If you operate the lift pump's priming lever and little or nothing happens, and the lever offers little or no resistance to movement, its operating cam has probably stopped, so that the diaphragm can't move. Give the engine a 'blip' on the starter to move the cam to a different position.

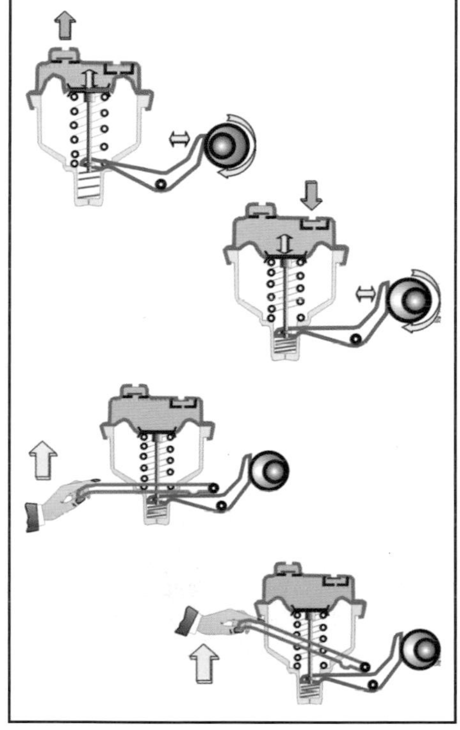

The fuel cut-off can be operated by one of two methods.

- Mechanically, usually by means of a cable and pull handle. Pulling the handle operates the stop lever on the engine.

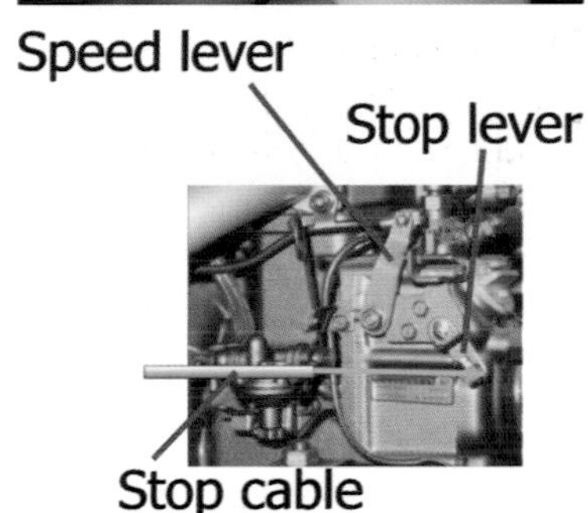

- Electrically using a solenoid operated by either the 'ignition' key or a stop button.

On a marine engine, unlike a car, the electrical stop solenoid is powered to stop. It doesn't need a 12 volt supply to keep the engine running. However, where the engine is part of a 'generator set' the stop solenoid is powered to run, so that in the event of a fault, such as low oil pressure, the engine is shut down automatically by removing the 12 volt supply.

BIOLOGICAL CONTAMINATION OF FUEL

Various forms of fungal and algal organisms can contaminate diesel fuel. In order that the organisms can grow, thrive and become a problem, they need two things:

- water in the fuel tank;
- a high enough temperature.

Diesel 'bug' will grow only at the interface of the water at the bottom of the tank and the fuel above it. Without water in the tank, the bug will not grow. If there is water in the tank AND you have contamination, only once the temperature is sufficiently high will significant growth occur. The ideal temperature for growth is between 15 and 35 °C, with rapid growth at 35 °C. The return fuel from the fuel injection pump is warm and will raise the temperature of fuel in the tank.

Water in your fuel tank will come from condensation in your, or the supplier's, tank or from water entering through your, or the supplier's, fuel filler, through bad design and poor sealing.

The bug will originally be airborne and there's quite a number of different species.

Some bugs produce sulphuric acid and all will cause 'sludge' in the tank. The acid causes corrosion in the fuel system and the sludge blocks the filters.

As far as you are concerned, your best line of defence is to ensure there's no water in the tank. This can be achieved by:

- buying your fuel from a respected source with a high turnover of fuel;

Tip

Get someone to operate the stop control while you look at the engine to see what moves. You will now know what stops the engine, so that in the event of a failure of the stop mechanism, you'll be able to operate it by hand. **Ensure that you don't get caught up in rotating machinery if you have to do this.**

- always keeping the tank topped up to prevent condensation (especially in the winter).

The second line of defence is a water separating prefilter (primary filter) with a transparent bowl. This allows you to inspect the bowl daily for the presence of water (and dirt).

Biocides

- Bacterial contamination can be killed with a biocide, but the dead 'bodies' and sludge can still cause corrosion and filter blockage.
- There are a number of different biocides available.
- If a biocide is used at every refuelling, the bugs will become resistant and greater concentrations will be needed. Eventually, it's possible that complete resistance will occur.
- Use of a biocide should be restricted to an actual case of contamination, it should not be used as a prophylactic.
- Biocides are hazardous to your health.

Water removers

- These substances take the water into suspension so that the water is removed from the tank as fuel is used.
- They are used in the road transport industry to good effect.
- In a leisure boat which is used infrequently, fuel, with its cargo of water, will remain in the fuel injection pump for long periods of time. I believe this is a cause of expensive corrosion in the pump and I don't recommend their use for leisure marine fuel tanks.

Magnetic fuel conditioners

I have seen no scientific proof of the effectiveness of magnetic (or ionising) devices.

Enzyme fuel additives

- Soltron, an enzyme additive, kills 99% of the biological contaminants and, over a succession of doses, will kill all the bugs.

- Soltron will remove biological sludge from a fuel tank.
- Soltron is not a biocide and can be used safely at each refuelling.
- Soltron will enhance combustion and increase fuel efficiency by around 5% in our type of use; it is thus cost neutral.
- I have seen a large amount of independent laboratory data on this product, all of which confirm the manufacturer's claims.

A bad case of diesel bug contamination, as illustrated, can be treated successfully using this product. After 120 hours running, with no action other than the use of Soltron at each refuelling, the filter is clean.

WINTERISATION OF THE FUEL SYSTEM

If the engine is going to be out of use for some time, then the fuel system needs to be *laid-up*. Usually, this lay-up procedure is combined with the annual service on most leisure boats, because the number of hours that the engine is run is relatively low.

1. Fill the fuel tank.
2. Change the filters and clean their cases/bowls.
3. Bleed the fuel system.
4. Start and run the engine.

The Engine Cooling System

Only about 50% of the energy in the fuel is converted into engine power! The other 50% is wasted as heat energy, about half going into the exhaust and the other half needing to be removed by the engine's cooling system.

AIR COOLING

Air cooling has a number of advantages, amongst which are lighter weight and quicker warm-up, but where installed in an enclosed engine compartment, as on most boats, getting the cooling air to and from the engine raises considerable problems. Noise is also a factor as is the uncooled exhaust pipe. Leisure boats rarely have air-cooled engines these days.

WATER COOLING

There are two forms of water cooling:

- *Direct (raw water) cooling.* The engine is cooled by the water in which the boat floats. So, if you are on the sea, seawater circulates through the engine's cooling system.

- *Indirect cooling.* Freshwater circulates around the engine's cooling system and is cooled by a heat exchanger. Usually seawater (or river water) flows through the heat exchanger to remove the heat. On some boats, a keel cooler, mounted outside the hull and immersed in the sea or river water, carries the heat away. On steel-hulled boats, the 'keel' cooler is welded to the inside of the hull plating and the plating itself transfers the heat to the water outside.

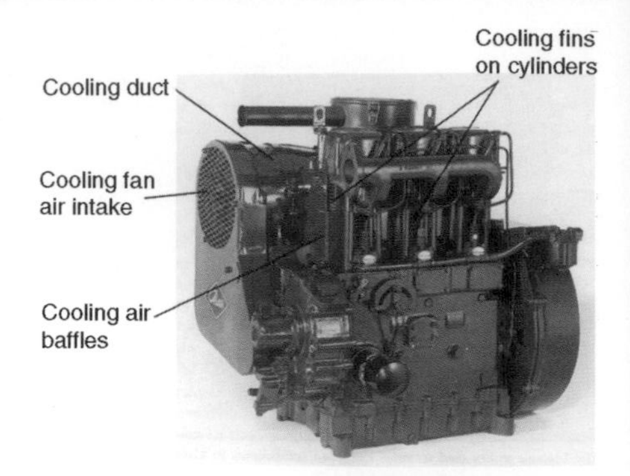

Air-cooled diesel engine

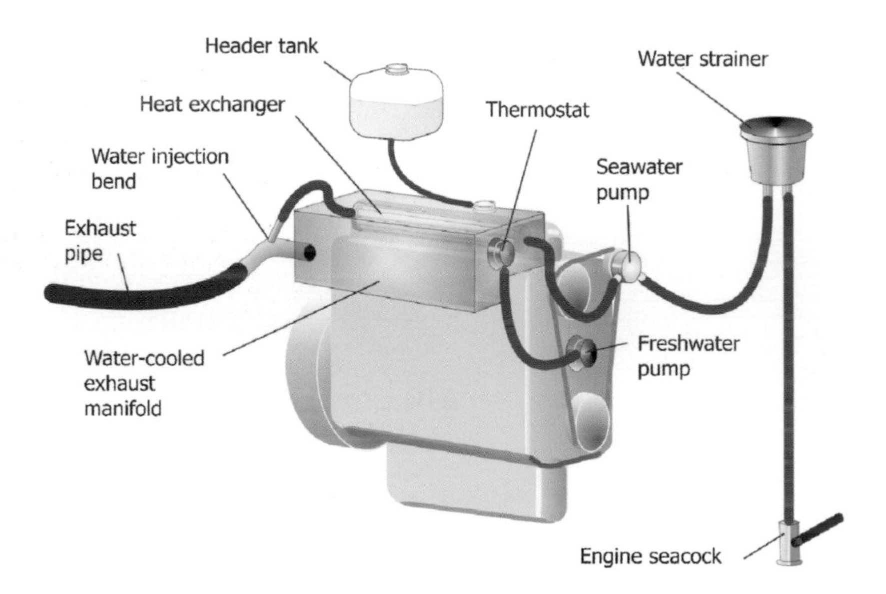

Engine cooling system

Water-cooled exhaust

Most leisure boats have rubber exhaust pipes and these must be water cooled so that they do not burn. Sailing boats normally have a 'swan neck' in the exhaust pipe to prevent a following sea flowing up the exhaust pipe and into the engine when the boat is sailing.

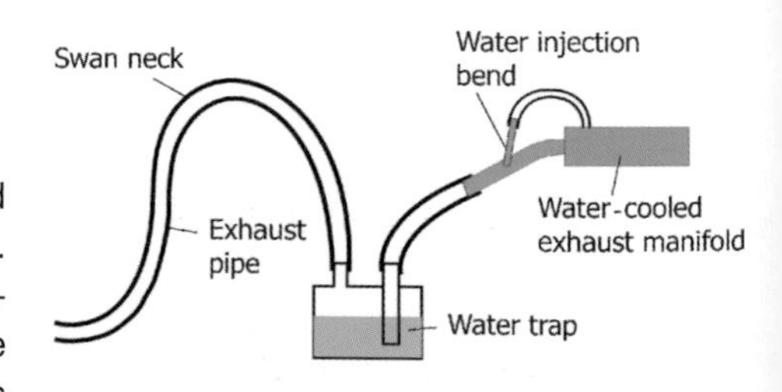

Sailing boat exhaust system

Direct cooling

With only one water pump and no heat exchanger, this is the simplest form of water cooling, but suffers from a major drawback. In the case of seawater being used for cooling, corrosion of the internal waterways will occur, and to reduce this to acceptable limits, the running temperature of the coolant is kept below 60 °C. This means that the engine is running at too cool a temperature for best efficiency, which is normally achieved much closer to 100 °C.

- The seawater pump sucks water through the engine cooling seawater cock and the water strainer. The strainer prevents weed, etc. from passing into the cooling system.

- Seawater is then directed to the thermostat.
  - If the engine is cold, water flows directly via the thermostat bypass to the water injection bend, without flowing though the engine's cooling ways.
  - Once the engine starts to warm up, the thermostat starts to open, allowing some of the cooling water to flow round the engine before rejoining the flow to the injection bend.

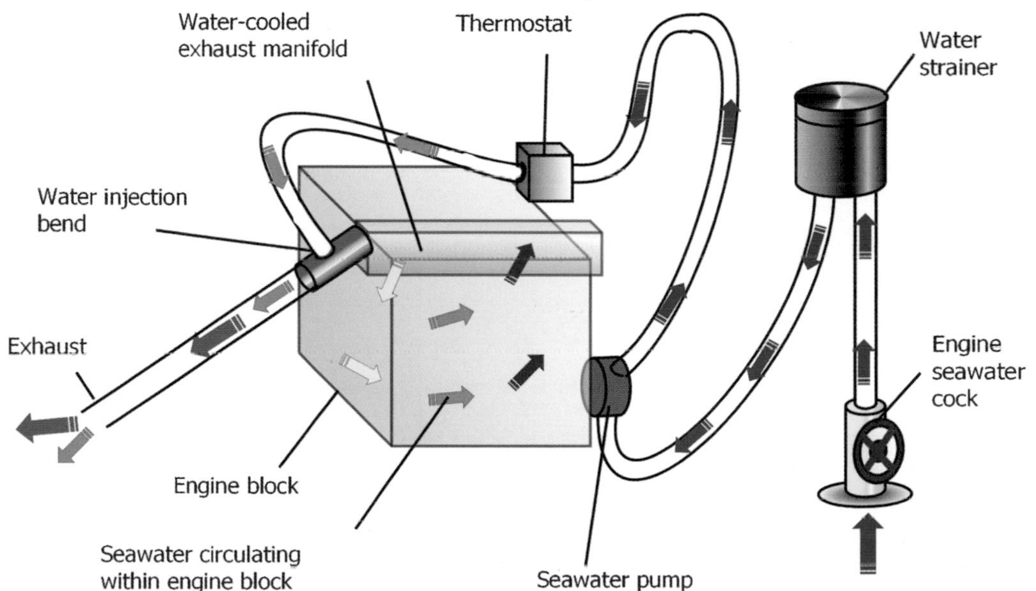

Direct engine cooling system

- The thermostat will open sufficiently to allow just the right amount of water to flow round the engine so that it runs at the desired temperature.
- Once the thermostat is fully open, all the water flows around the engine, no additional cooling will take place and the temperature will continue to rise, eventually setting off the over-temperature alarm.

If yours is a directly cooled engine using seawater, ensure that any cooling system anodes are checked at least annually.

Modern engines, unless designed specifically for seawater cooling, will not have direct cooling systems.

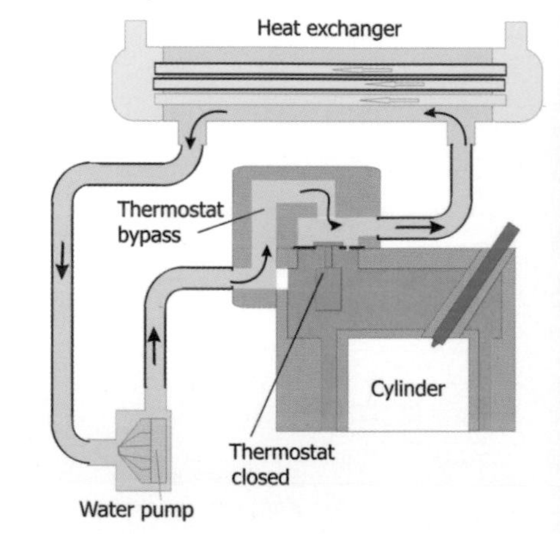

Cold engine, thermostat closed, all water bypasses engine

Indirect cooling

The seawater path:

- The seawater pump sucks water through the engine cooling sea water cock and the water strainer.
- The strainer prevents weed, etc. from passing into the cooling system.
- From the seawater pump, water flows through the tubes of the heat exchanger before being discharged directly into the exhaust pipe via the injection bend. It does not flow through any of the engine's cooling ways. On some engines the heat exchanger is mounted inside the coolant header tank, and sometimes it is mounted separately.

The freshwater path:

- Freshwater is circulated around the engine's cooling ways by means of a freshwater pump or circulator.
- The thermostat controls the amount of freshwater flowing through the heat exchanger.
- As the water temperature starts to rise, the thermostat begins to open, allowing some freshwater to pass through the heat exchanger, where it is cooled by the seawater before being returned to the engine.
- In normal operation, just sufficient freshwater is passed through the heat exchanger to maintain the required operating temperature.

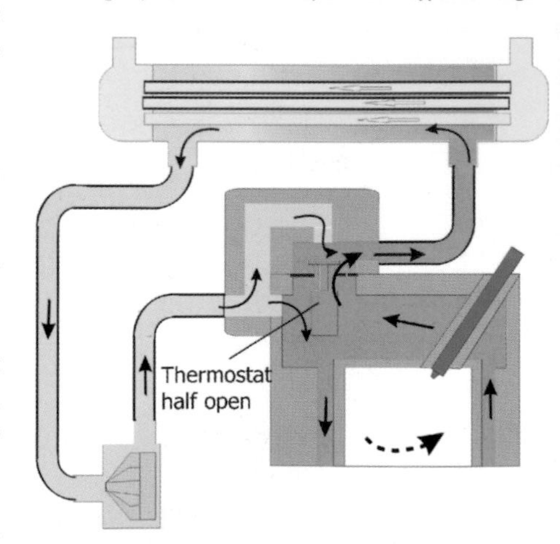

Thermostat half open. Some water bypasses the engine and some water flows through the engine to maintain the desired running temperature

- If the water gets too hot, the thermostat opens fully and all the water passes through the engine.

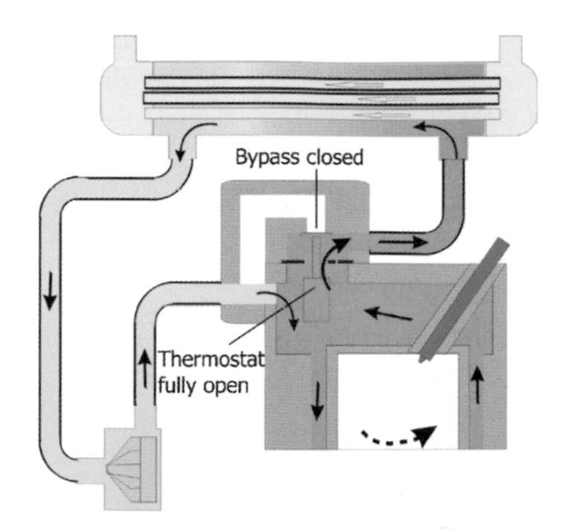

Thermostat fully open, blocking off bypass, so all water flows through engine

This is identical to the way in which water is directed through the radiator on a car.

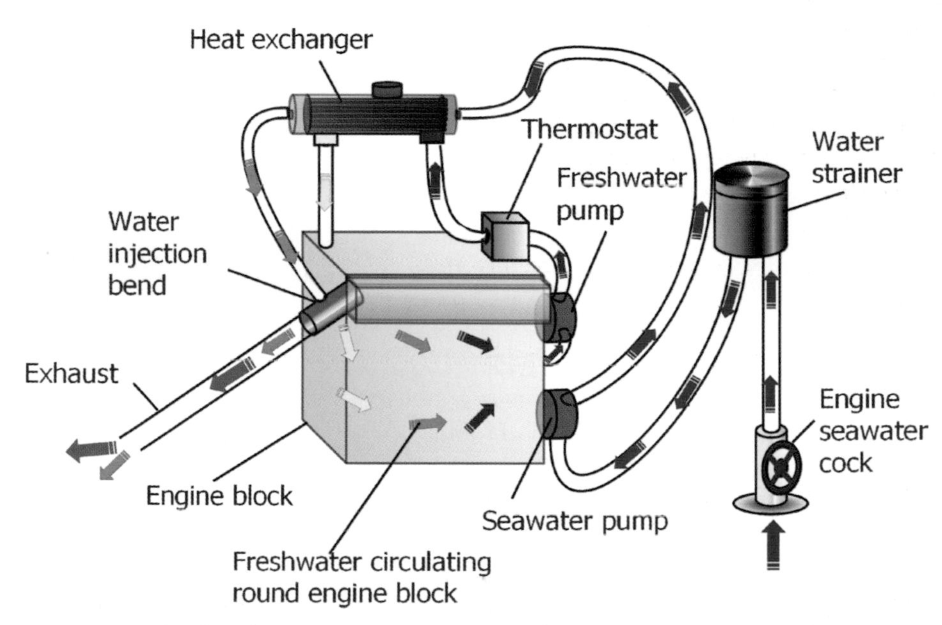

Indirect engine cooling system

Keel cooling

Some boats, especially those operating on canals, use *keel cooling.* This is a form of indirect cooling, where the heat exchanger is either welded to the steel hull of a boat, usually on the inside of the bottom of the hull, or in the form of pipes on the outside of the hull. In the latter case it is not confined to steel hulls.

The advantage is that no seawater pump is required, and if the heat exchanger is inside the hull, no sea cocks are needed either.

The disadvantage is that a dry exhaust is needed, so some means of insulating it is required.

WATER FLOW FAILURE

A failure in the water flow can have rapid and serious consequences, with damage occurring before the engine overheat warning is activated. Few manufacturers incorporate a rapid failure detection system as standard.

A blockage in the seawater intake will cause the impeller to run dry and it will break up due to overheating, often before the overheat warning is activated. Impeller debris can lodge in and block the heat exchanger of an indirect system or inside the waterways of a directly cooled engine.

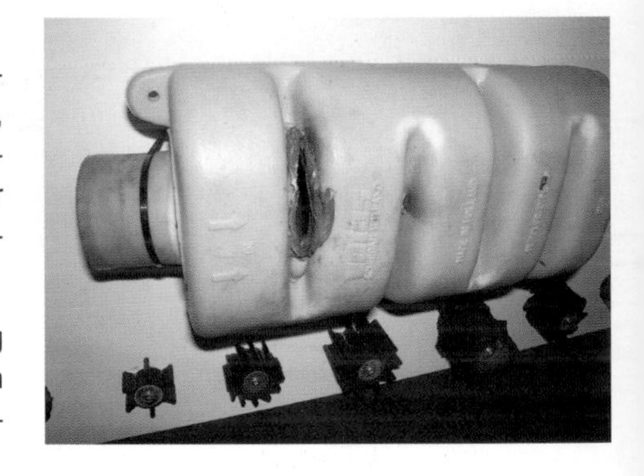

Failure of the water flow will rapidly cause overheating of the exhaust system, which can then burn through allowing exhaust gases into the boat's accommodation, possibly with fatal results.

SEAWATER (RAW WATER) PUMP

Almost all engines use a flexible impeller self-priming pump to make the seawater flow through the cooling system. The pump is driven either by an external belt drive or internally from the camshaft. A flexible impeller rotates within a tightly fitting enclosure. The front plate of the pump is removable so that the impeller may be examined or changed. It is normally simple to remove the pump for repair or exchange.

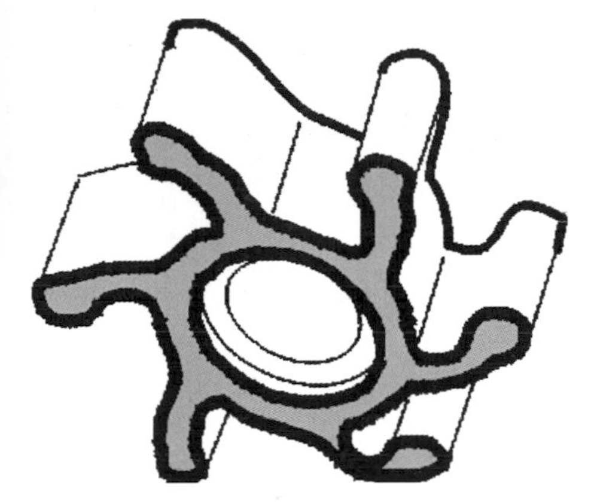

Principle of the impeller pump

- The pump uses a multi-bladed flexible impeller.
- Inside the pump body is a brass 'cam'.

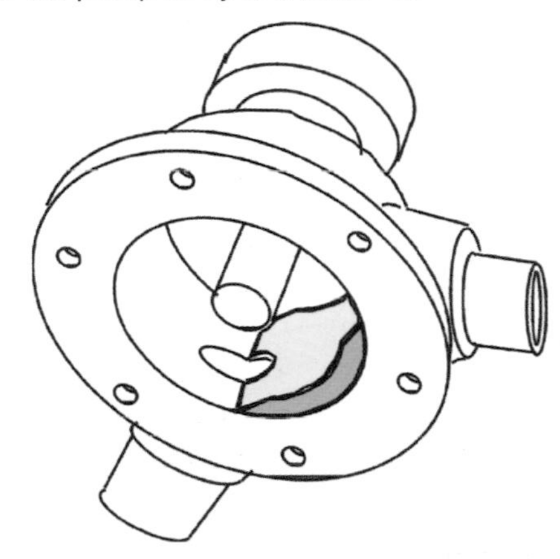

- The cam distorts the impeller blades as the impeller rotates.
- As the impeller rotates, it draws in water at the inlet, carries it round the pump body and forces it out of the pump outlet.
- The impeller is lubricated and cooled by the water being pumped.

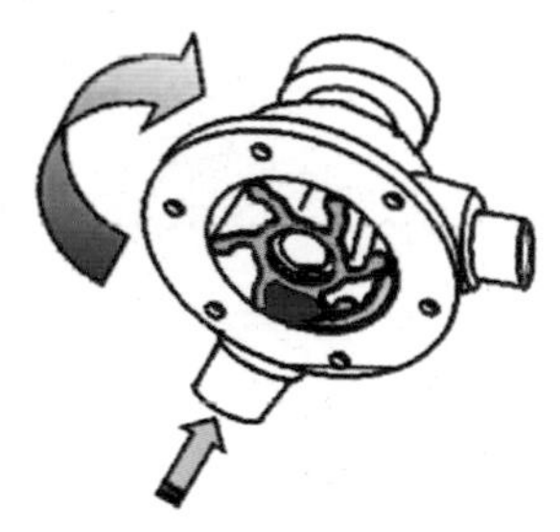

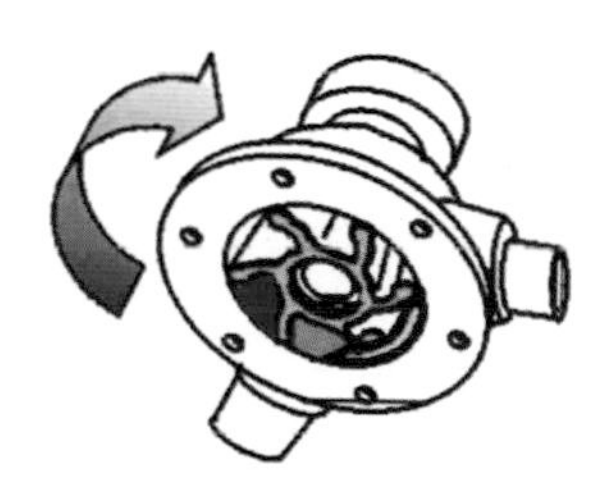

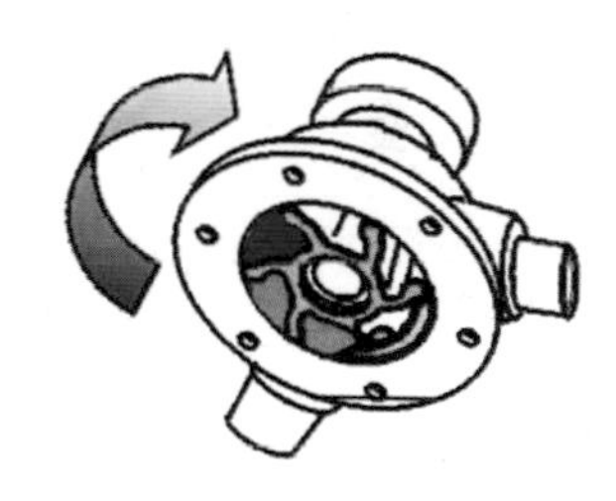

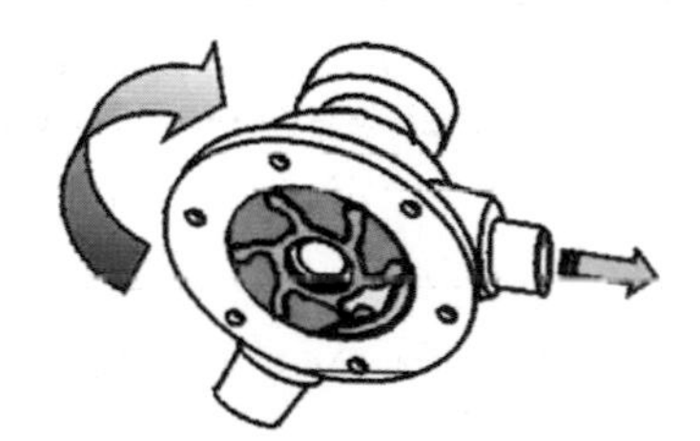

Changing the impeller

- Turn off the engine-cooling seawater cock.
- Undo the retaining screws of the pump face plate.
- Remove the face plate and gasket.

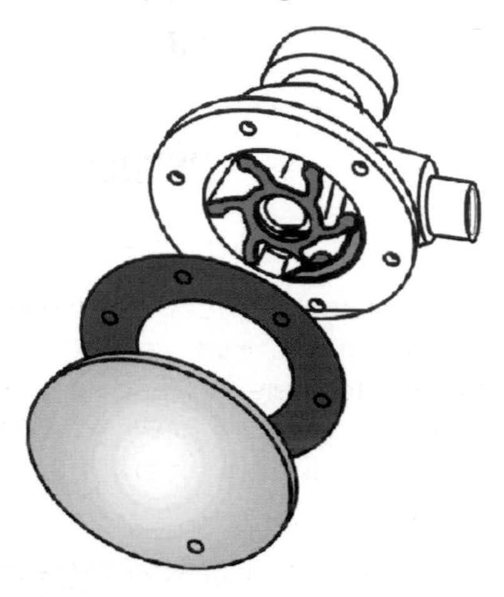

- Remove the impeller.
- The best tool for removing the impeller is a dedicated 'puller', and this is essential for large engines.
- Insert the new impeller.
- Replace the cover and gasket.
- Insert and tighten the screws (but don't over-tighten).
- Open the seacock.

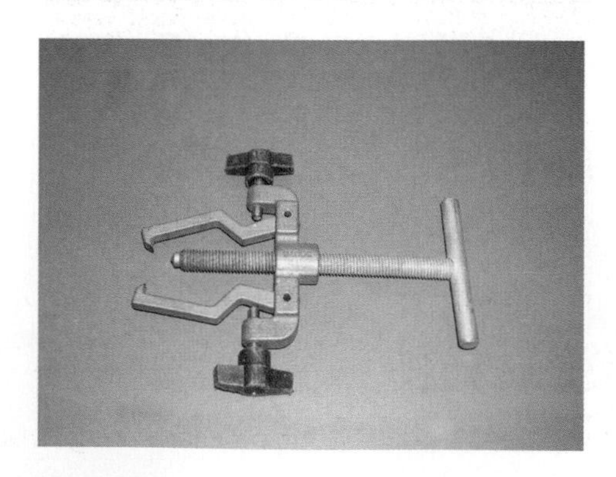

Tips

- Use a 'Speed Seal' from True Marine to make impeller changes much easier as the endplate can be removed without tools.

- Bind the impeller with a cable tie, rubber band or string to reduce its diameter so that it slips in easily.

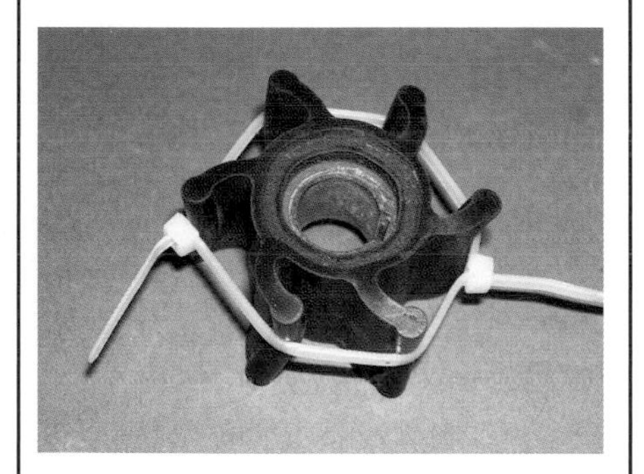

- The impeller always rotates the long way round from inlet to outlet, so make sure you insert the impeller with its blades trailing.

When you restart the engine, check the exhaust to make sure that cooling water is coming out, then check the pump face plate joint for leakage. If the face plate is hot to the touch, it's running dry. Make sure that you don't get caught up in any moving parts of the engine or shaft.

One yachtsman had a cooling water failure and, on removing the face plate and impeller, a small eel dropped out into the bilge! You never know what you might find.

Where's the seawater pump?

1. On many smaller engines, the seawater pump is located on the front of the engine, where, generally, it's accessible. However, that's not always the case.

2. Older Volvo Pentas have them above the gearbox.

3. Follow the pipe from the seawater strainer and you'll come to the pump.

Causes of pump failure

1. If the pump runs dry, it will overheat and the blades will disintegrate. This is the most common cause of failure.

2. Pieces of blade can lodge in the cooling ways causing a partial blockage, which will then contribute to further overheating. In seawater-cooled engines (direct cooling), the debris could end up almost anywhere and may or may not cause any further problem. Back flushing the cooling system may cause the debris to return to the pump body, so if you are concerned, you may well have to consult an engineer.

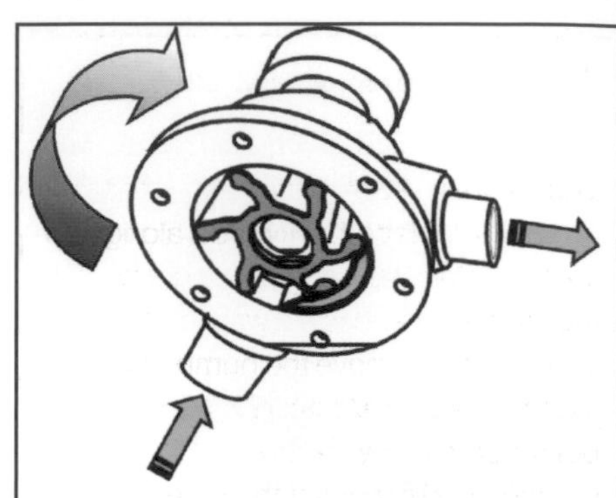

- If the gasket is unserviceable, use non-hardening gasket sealant.

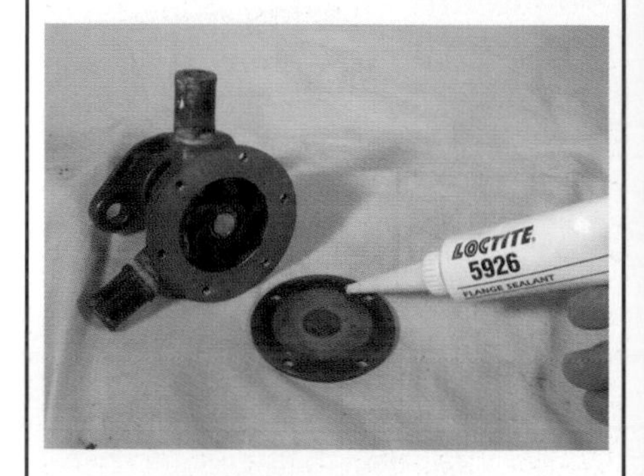

- For bigger pumps, you'll need a proper impeller puller tool.

Tip – Yanmar 2GM and 3GM engines

The Yanmar 2GM and 3GM, although on the front, face backwards alongside the front starboard engine mounting. On the Yanmars the easiest way to get at the impeller is to remove the pump first. This isn't as bad as it sounds, as only two bolts need to be removed and there's no need to disconnect the pipes.

3. With freshwater-cooled engines (indirect cooling), debris will go directly to the heat exchanger and either lodge there or pass right through into the exhaust. Cleaning the heat exchanger tubes is a comparatively simple task.

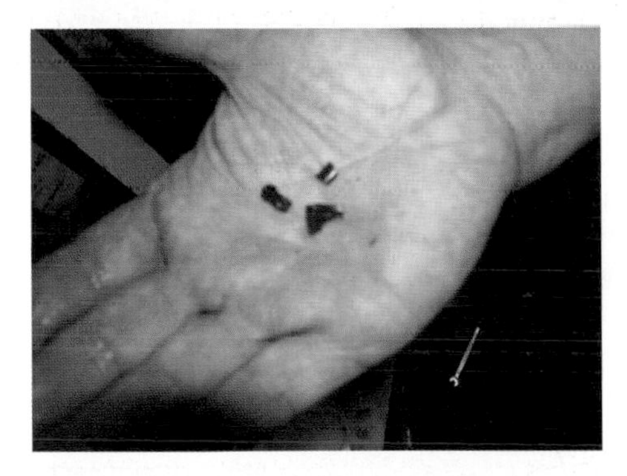

4. Pump impellers sometimes become unbonded from their central metal boss. The impeller then remains stationary in the pump while the boss drives round. When you remove the face plate this is not at all obvious, so when you extract the impeller, try turning the impeller while you firmly grip the boss to check the bonding. Several batches of impellers suffered from this problem, so this can occur very quickly, even with a new one.

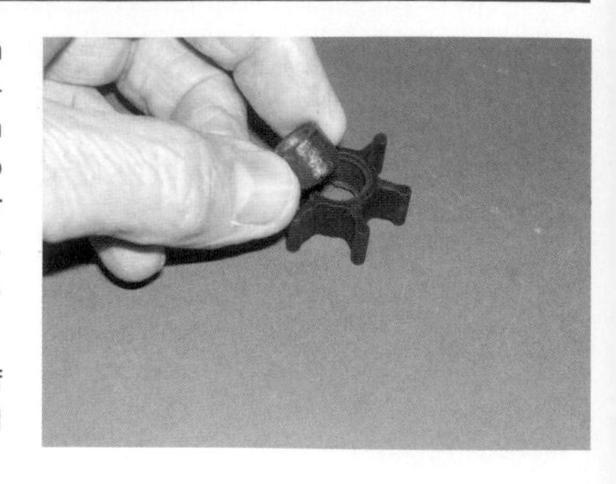

5. Wear of the pump cam will cause a reduced rate of flow. This reduced flow may be insufficient to cool the engine and overheating can then occur.

6. An even rarer problem is for the screw holding the wedge–shaped cam inside the pump case to shear. This cam then rotates with the impeller rather than staying stationary, so check that the cam is between the inlet and the outlet the 'short way round'. If the screw has sheared you'll probably have to remove the pump to replace the screw. The replacement must not be so long as to stand proud of the cam or the impeller will be damaged.

7. If there is significant wear on the front plate, as a temporary measure it can be reversed so that the front side faces the impeller. Any inscription will not affect the pump's performance.

Servicing the seawater pump

Servicing is usually required when the water and oil seals start to leak. The oil seal is usually no problem, but the water seal will wear due to silt carried in the water and water will start to drip from the 'tell tale'.

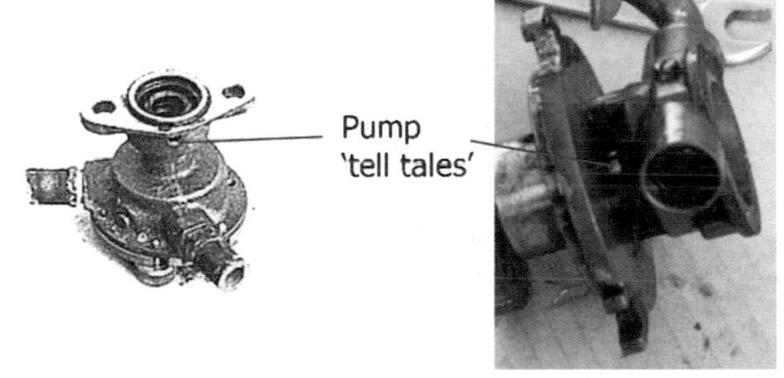

Pump tell tales (drip holes)

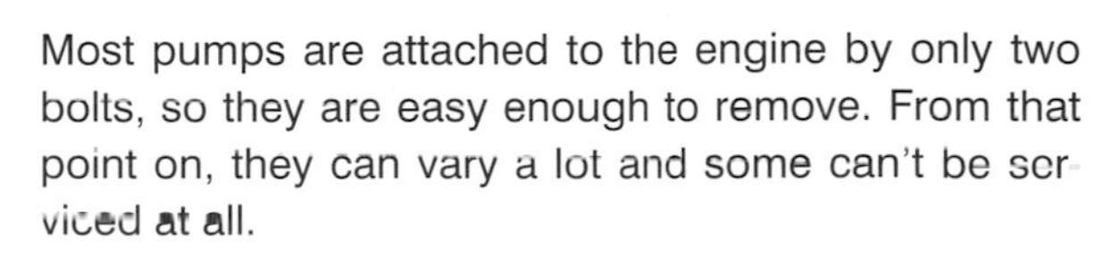
Most pumps are attached to the engine by only two bolts, so they are easy enough to remove. From that point on, they can vary a lot and some can't be serviced at all.

A gear-driven seawater pump is shown in the first photograph.

The Volvo Penta 2000 series pump comes off complete with its nylon drive gearwheel.

1. Remove the impeller.

Impeller removed from pump

Pump removed from Volvo Penta 2000 series

2. Remove the circlips and the gearwheel.

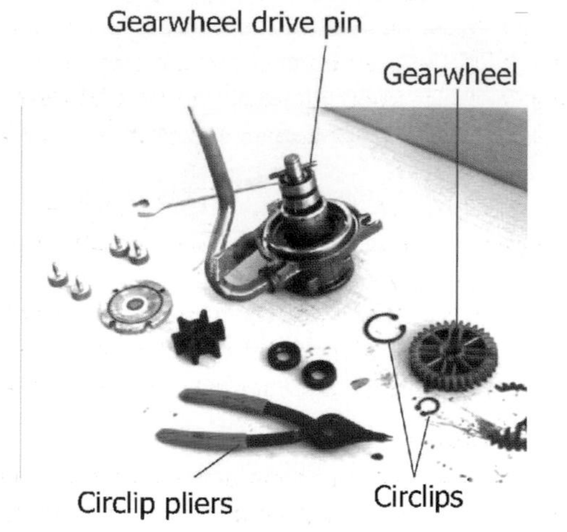

Remove gearwheel

3. Remove the shaft, bearings and seals.

Shaft and bearings removed

Using a 'drift' to punch
out the bearings and seals

4. The bearings and seals may need to be driven out with a drift.
5. Examine the shaft for wear. If you leave replacing the seals too long, a groove will be worn in the shaft so you'll have to replace that as well.

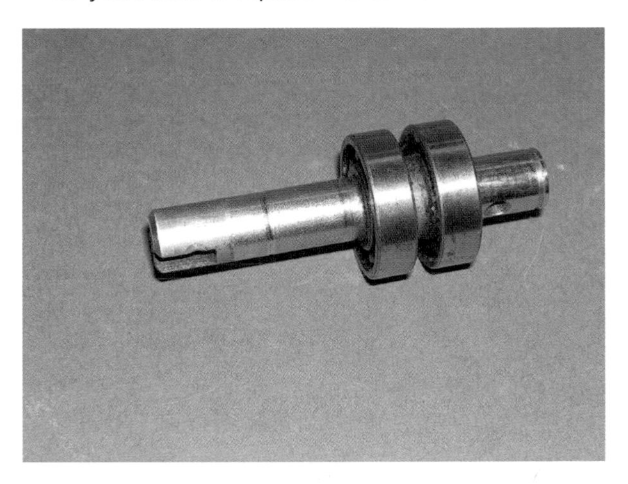

6. Replace the seals (and shaft and bearings if necessary). Reassemble in the reverse order.

SABB WATER PUMP

Sabb engines use a diaphragm pump, which has no impeller. It's a single pump on seawater–cooled engines and a double pump in the case of freshwater cooling. Sabb instruction books have very full details of the operation of the pump and routine maintenance.

FRESHWATER CIRCULATING PUMP

The freshwater pump has a rotor and no other moving parts. The pump rotor itself can't fail, but its bearings and seals can. If the pump is belt-driven, then slackening it off when the boat is laid up will prolong its life; however, most are driven internally and you have no access.

The freshwater pump is a circulator using a centrifugal impeller. It is normally mounted on the front of the engine, and may be mounted externally, where it is accessible, or it may be mounted inside the engine casing, where it is inaccessible. You will need to look at the engine's workshop manual for information on replacing bearings and seals, as they all differ so much.

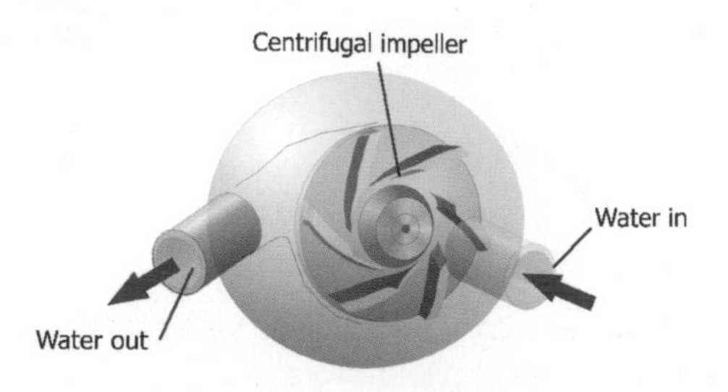

Freshwater pump

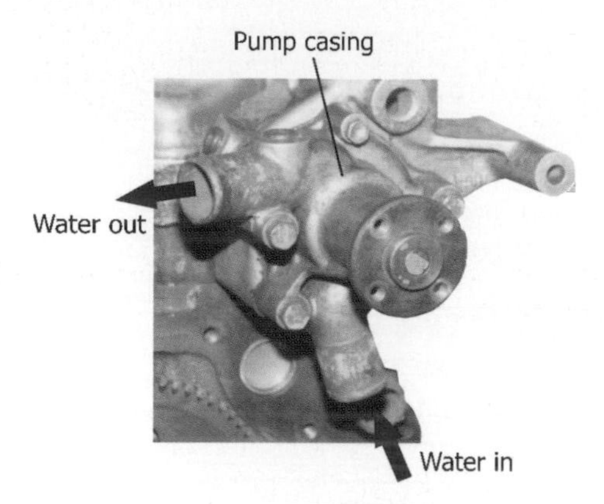

Freshwater pump

HEAT EXCHANGERS

Indirect cooling systems require a unit to transfer the 'engine' heat in the freshwater system to the 'raw water'. This unit is called a *heat exchanger* and consists of a bundle of copper tubes, called a *tube stack*, through which seawater flows.

The tube stack is enclosed in a container through which the fluid to be cooled circulates, heat being transferred from one to the other by conduction through the walls of the copper tubes.

Tips

- If you can't get the bearings out, have a word with your service agent who might remove them with his hydraulic press, especially if you buy the new bits from him.
- The bearings are likely to be an interference fit onto the shaft. Try putting the shaft in the freezer for a few hours and the bearings in a hot oven for a bit. If you then work quickly it's likely you will get them on.
- The oil and water seals are usually exactly the same. The hollow 'C'-shaped side faces the liquid you are trying to seal against.

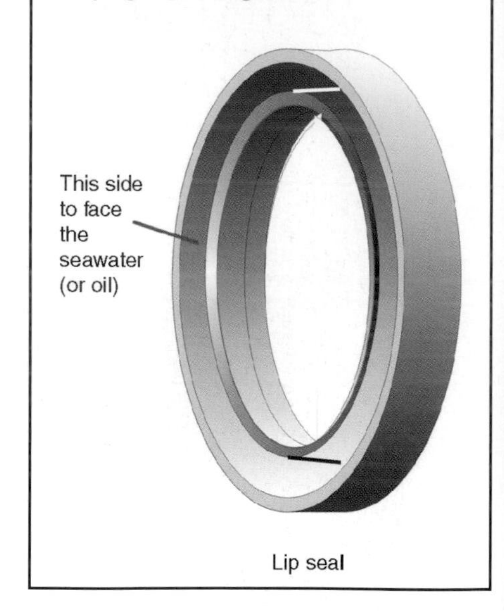

Lip seal

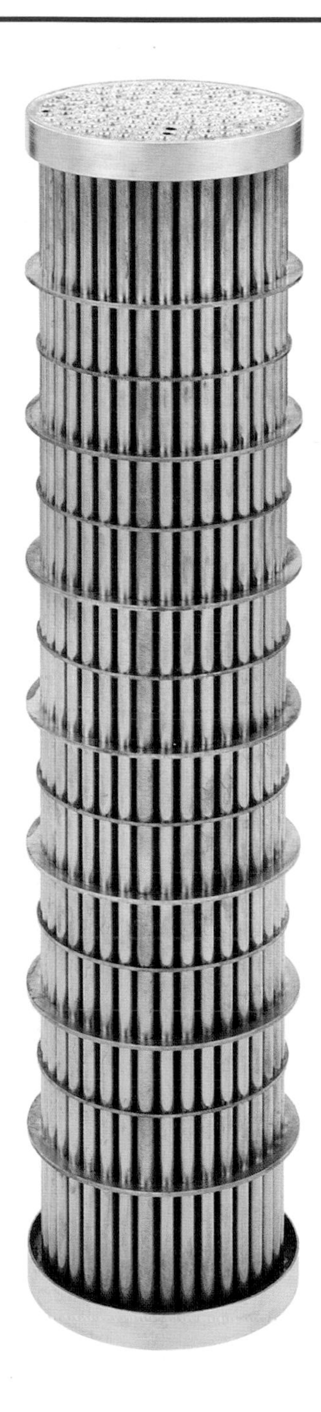

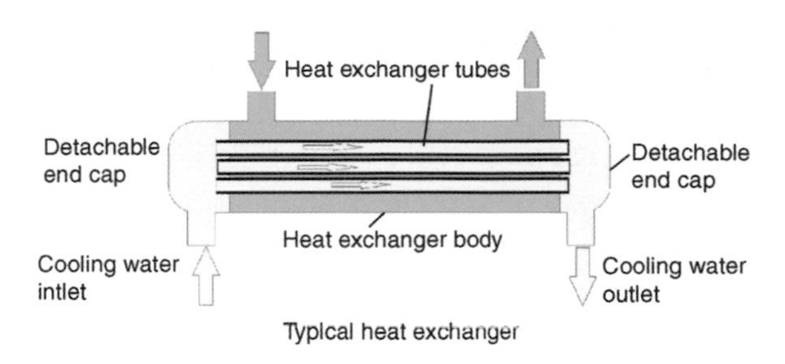

Typical heat exchanger

- The heat exchanger may be a self-contained unit mounted externally from the engine.

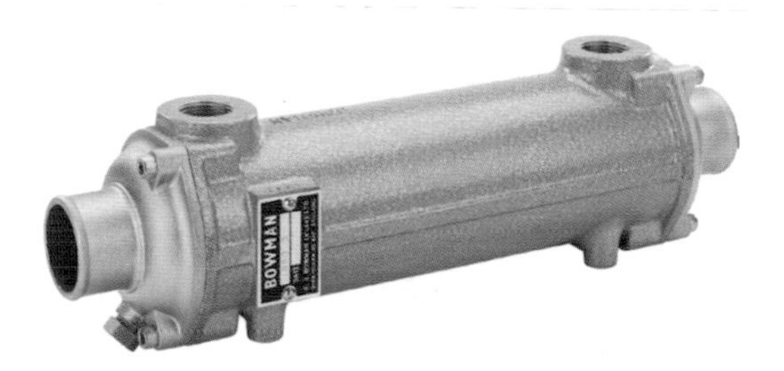

- The tube stack may be mounted within the cooling water header tank.

- In the case of a turbo-charger inter-cooler, the hot air flows through the tubes and the cooling seawater circulates round them.

Cooling systems can become quite complicated, especially on engines of high power. The outside of the engine can be encumbered with so much equipment that it isn't easy to sort out exactly what various things are. No different really from the modern car.

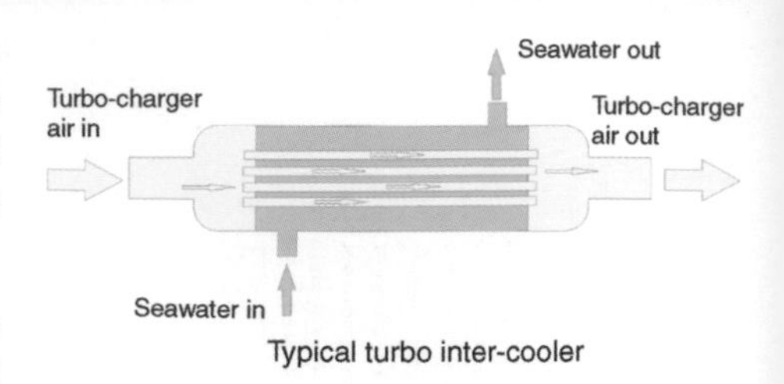

Typical turbo inter-cooler

The following parts may be incorporated into the system in addition to the engine-cooling water heat exchanger:

- turbo-charger inter-cooler;
- engine oil cooler;
- gearbox oil cooler.

This can add up to four heat exchangers, all in the seawater cooling circuit.

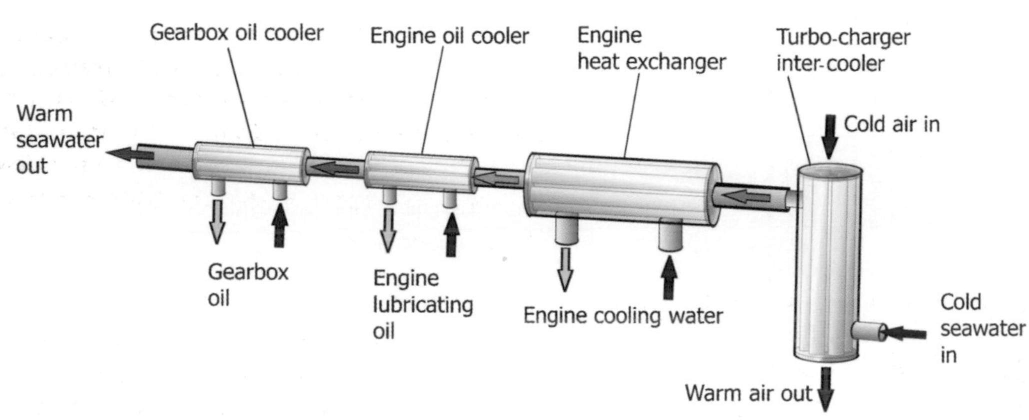

Multiple seawater-cooled heat exchangers

Heat exchanger problems

Blocked tubes

The inside of the tubes can become blocked due to a build up of deposits from the saltwater, reducing the cooling effect of the heat exchanger. Ideally, the tube stack should be removed and immersed in a non–caustic solvent approved by the heat exchanger manufacturer. Designs of heat exchangers vary and it may be advisable to follow the engine workshop manual or the heat exchanger manual before you take the heat exchanger apart.

1. Close the engine-cooling seawater cock.
2. Remove the face plate of the seawater pump to drain the seawater.
3. Partially drain the freshwater-cooling system by means of a drain tap or by removing a freshwater pipe from a fitting below the heat exchanger.
4. Remove the end plates from the heat exchanger.
5. Pull the tube stack clear of the heat exchanger, taking care not to damage any seals.
6. Immerse the tube stack in the solvent, observing the instructions carefully.
7. Reassemble in the reverse order, ideally using new seals.

Alternatively, the tubes can be cleaned by passing a length of steel rod down the tube to dislodge all the fouling. *It is important that the rod is 'deburred' at its end and that you don't dig the end of the rod into the soft copper wall of the tube.* Push the rod down the tube in the opposite direction to the flow of the water.

Seawater leaking into the freshwater system

Seawater can leak past a seal into the freshwater. This is most likely if a seal is damaged on reassembly or due to loose jubilee clips where a rubber *boot* type end-cap is used. The symptoms of this are continuous overflow from the header tank filler cap overflow with no necessity to top up the freshwater. This results in the freshwater becoming contaminated with seawater, giving rise to internal corrosion.

Combined header tank and heat exchanger with rubber end covers

Check the tightness of the jubilee clips on the end covers regularly. If the smaller of the two jubilee clips becomes loose, seawater can be forced past the rubber boot into the freshwater, contaminating it.

Heat exchanger anodes

See later in this chapter for details of anodes, which must be checked regularly.

SYPHON BREAK

Seawater in the working parts of the engine

On sailing boats the exhaust pipe is swept up into a 'swan neck' to stop following waves flooding the exhaust when the engine isn't running. Exhaust gas normally drives water out of the system up over the swan neck, but of course if the engine isn't running and seawater is entering the system, water will accumulate in the exhaust. Eventually, there will be enough to enter the exhaust manifold and into any cylinder that has its exhaust valve open.

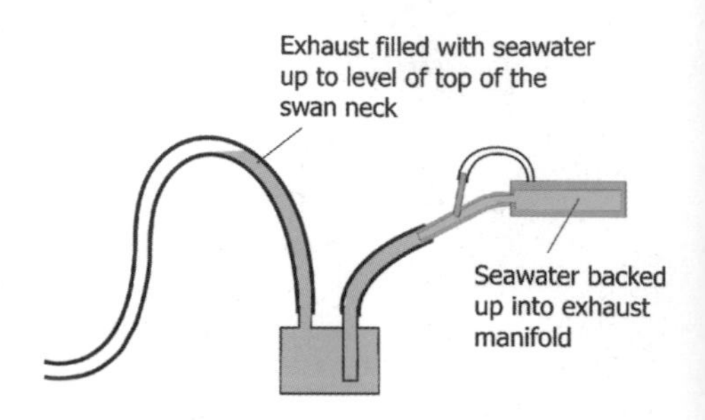

Flooded exhaust system

- Water can enter the exhaust when water siphons past the water pump, because the exhaust injection bend is below the water line.
- On a sailing boat and some motor boats, water can be trapped in the exhaust system if you continue to turn the engine on the starter motor and the engine doesn't fire.
- Water entering the working parts will contaminate the oil, causing serious corrosion.
- Water is incompressible and can shatter the piston or bend a connecting rod.

Syphon break system

Any water-cooled engine mounted fairly low in a sailing boat is at risk from seawater entering the working parts of the engine. Water can syphon into the exhaust pipe via the open seawater cock because the water injection point into the exhaust is below the waterline. To prevent this happening, a syphon break should be fitted.

If one is needed, the builder will probably have installed one. You may not know it's there, as it needs to be fitted at least 150 mm above the waterline, so it's often out of sight.

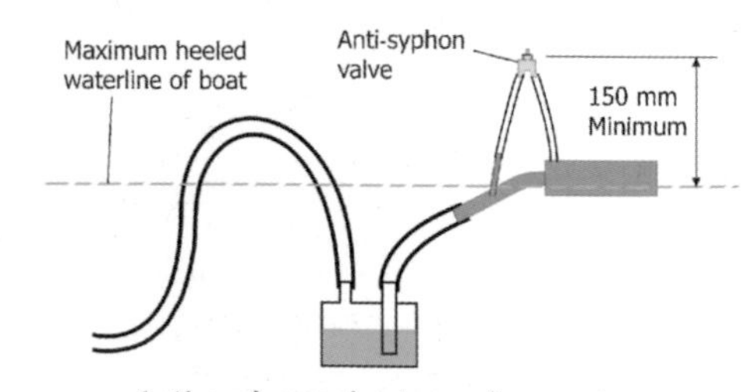

Anti-syphon valve in cooling system

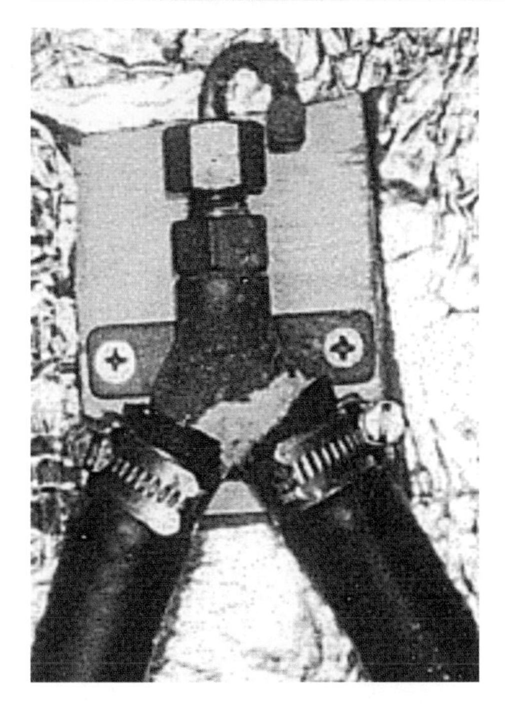

To find out if you have one, look to see if the seawater supply pipe from the pump runs upwards above the engine and then returns down before going into the cooling system. You might instead find it doing the same thing, but after it leaves the engine prior to entering the exhaust.

A rubber valve in the anti-syphon valve is pushed closed by the pressure from the flow of seawater from the seawater pump. When the engine stops, atmospheric pressure opens the rubber valve allowing air into the pipe, breaking the syphon.

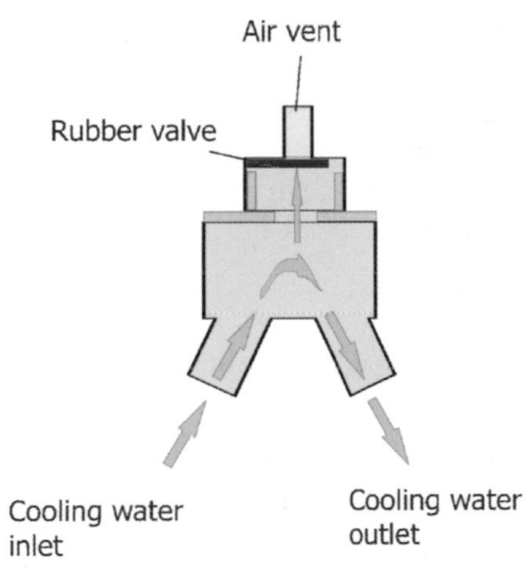

Anti-syphon valve engine running

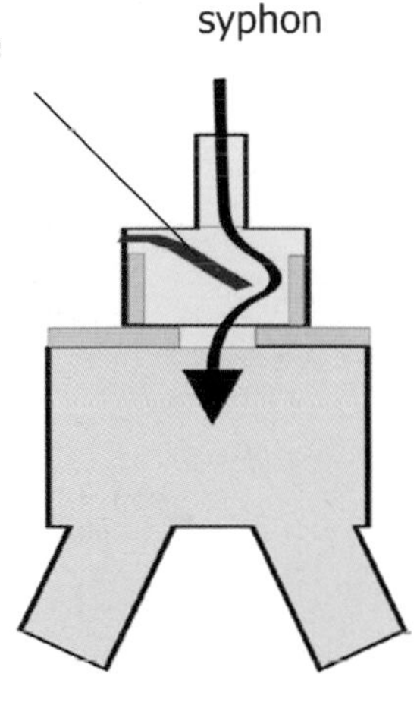

Anti-syphon valve engine stopped

The ideal system uses no valve, but has a small–diameter tube draining overboard well above any possible heeled waterline. This is a 'fail safe' system and requires no maintenance except to observe that there's a flow of water out of it when the engine's running.

When the engine is running, the syphon break allows a small flow of water to be vented overboard so that you can see the system is not blocked (and indeed that the pump is functioning). When the engine stops, air is admitted to the system, breaking the syphon – you can often hear the air being sucked in as you stop the engine.

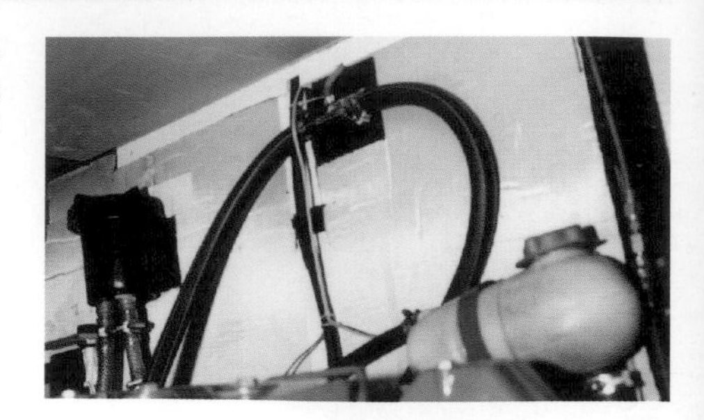

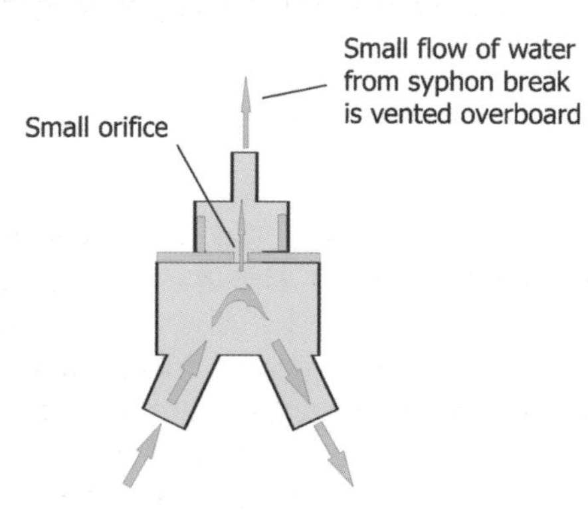

Syphon break
engine running

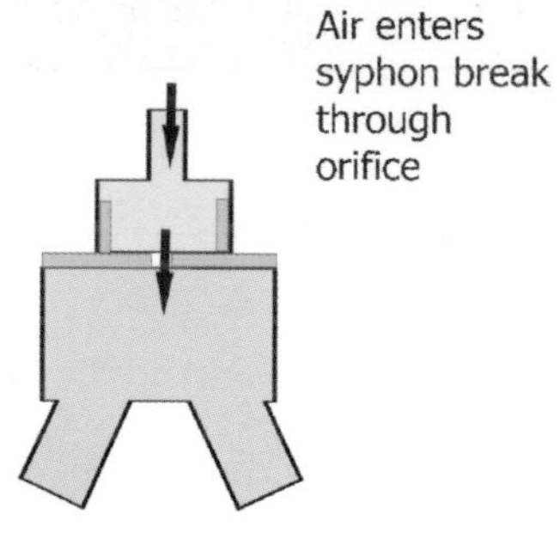

Syphon break
engine stopped

Cleaning the anti-syphon valve

If there is an anti-syphon valve, it must be cleaned annually. You know it's a valve, because there's no tube running overboard from it, although there may be a short length dropping down towards the bilge.

Tip

Volvo Penta anti-syphon valve

This unit must be held upside-down when you reassemble it. It's often much easier to remove this valve from the bulkhead when you service it, unless it's very accessible.

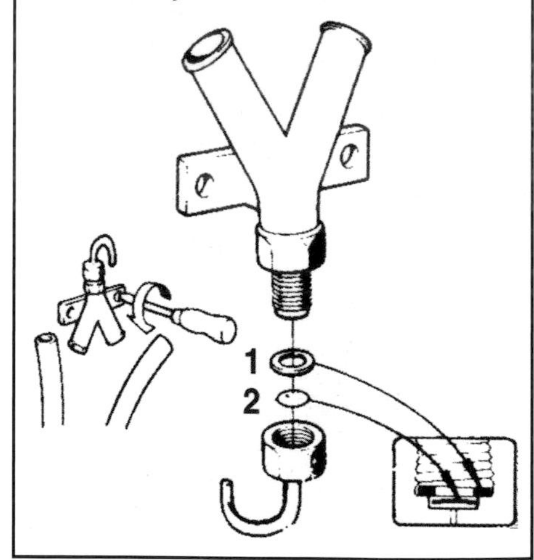

1. Remove the top from the valve assembly.
2. Carefully remove the valve and clean it and the seating.
3. Check that the vent pipe is clear by blowing through it.
4. Carefully reassemble the unit.

THE THERMOSTAT

The thermostat is the device that controls the running temperature of the engine's cooling system. A temperature-sensitive capsule is positioned in the coolant in the engine block. Expansion of the capsule due to rising coolant temperature causes a valve to open, allowing cold coolant to flow into the system to control the running temperature. The thermostat has a set temperature at which the valve starts to open, and it becomes fully open at another set temperature. These temperatures vary from engine to engine and also from direct- to indirect-cooled engines. A very small bleed hole always allows a small amount of cold coolant to flow through the thermostat, even when the valve is closed.

Operation of the thermostat

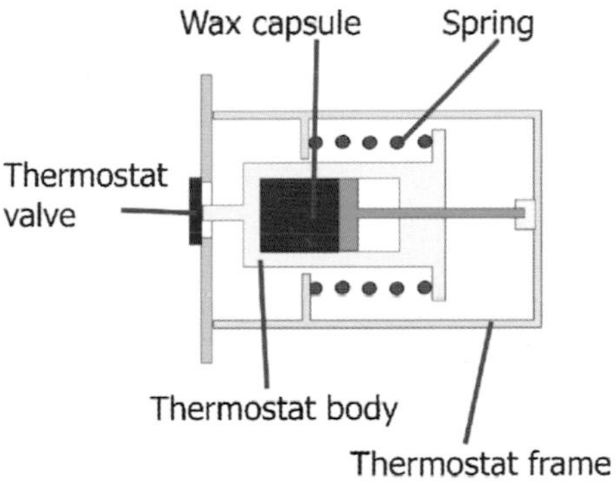

Wax capsule
Spring
Thermostat valve
Thermostat body
Thermostat frame

Thermostat closed

Wax expanding pushing piston against frame and forcing thermostat body to the left

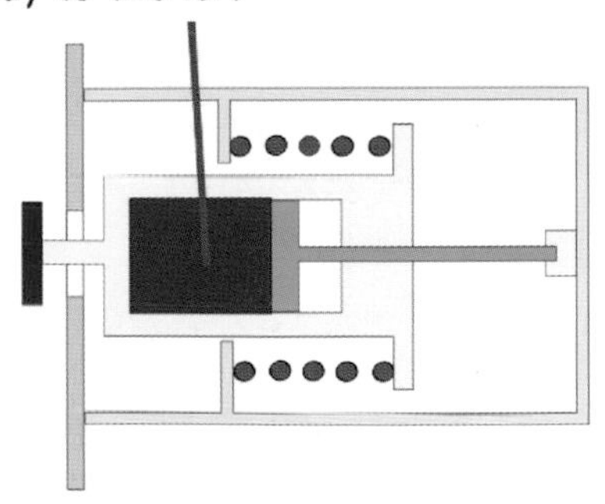

Thermostat half open

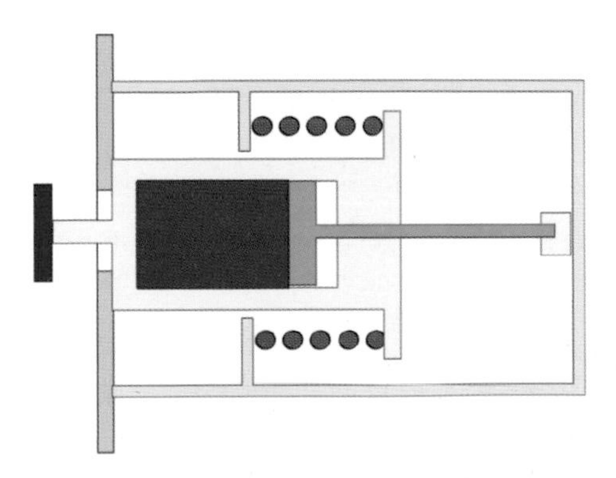

Thermostat fully open

Direct (seawater) cooled engines

Thermostats intended for use in seawater are generally set to operate at around 55 °C to reduce the effects of corrosion.

- When cold, all the incoming cold water bypasses the engine and is injected directly into the exhaust.
- When the thermostat starts to open, some cold water is allowed to flow into the engine to mix with the hot water.
- At the designed running temperature, the thermostat valve continually adjusts its position to allow just sufficient cold water into the engine to maintain the desired temperature.
- If the engine overheats, the valve opens fully to force all the cold water through the engine.
- After this point is reached, any further overheating is uncontrolled by the thermostat, which is now fully open.

Indirect (freshwater) cooled engines

Thermostats intended for use in freshwater-cooled engines are generally set to operate at around 90 °C to maximise engine efficiency, as corrosion is not a problem.

- When cold, no circulation of engine coolant occurs through the heat exchanger.
- When the temperature approaches the designed running temperature, the thermostat valve starts to open, allowing some engine coolant to flow through the heat exchanger.
- At the designed running temperature, the thermostat adjusts its position continually to maintain the desired running temperature.
- If the engine overheats, the valve is fully open and all the coolant is forced through the heat exchanger.
- After this point is reached, any further overheating is uncontrolled by the thermostat, which is now fully open.

Thermostat failure

Modern, wax-filled thermostats are generally reliable. If they do fail, they tend to do so in such a way that they will cause the engine to run too cool, rather than overheat. This has efficiency implications but will not cause the engine to overheat.

Older types, which are still to be found in many engines, have air-filled bellows, and when these fail they will normally cause the engine to overheat because the valve will stay closed.

Changing the thermostat

Where to find it

Look for the thermostat housing (often at the end of a flexible pipe) just after the seawater pump, it's normally held in place by a couple of bolts.

How to do it

1. Turn off the cooling seawater cock.
2. Loosen the seawater pump face plate to drain water from the thermostat housing. This is not essential, but stops seawater flooding over the engine.
3. Loosen the pipe clamps on any inlet and outlet pipes on the thermostat housing.
4. Remove the bolts attaching the thermostat housing.

5. Lift the housing clear.
6. Lift out the thermostat.

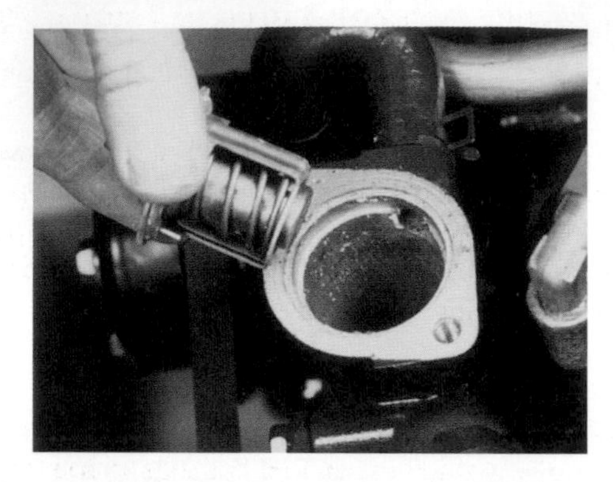

7. Test or renew the thermostat and replace it in the housing.
8. Reassemble in reverse order, renewing the gasket if necessary.
9. Open the cooling seawater cock.
10. Restart the engine, check the cooling water flow and check for leaks.
11. The housing will remain cool until the thermostat opens and then it will become hot to touch as hot water starts to flow inside it.
12. Recheck cooling system symptoms.

Running with the thermostat removed

DON'T!

Many sources still say that you can run the engine with the thermostat removed if the thermostat fails closed. In most cases you can't, because the big hole now introduced into the system will allow the cooling water to take the path of least resistance, which is normally straight out, without flowing round the cooling ways. As the temperature sensor is placed close to the thermostat, it now experiences only the cold water, so it will indicate a cool running engine while deep inside the water is boiling.

Tip – Testing a thermostat

The thermostat can be tested easily for correct operation. If you put it in a saucepan of water and heat it on the stove, it will be seen to open as the temperature rises. A thermometer will indicate correct operation, as the opening temperature will be stamped on the thermostat. In many cases the thermostat is changed unnecessarily in the hope that it will cure a problem, when a simple check would have revealed that it was, in fact, serviceable.

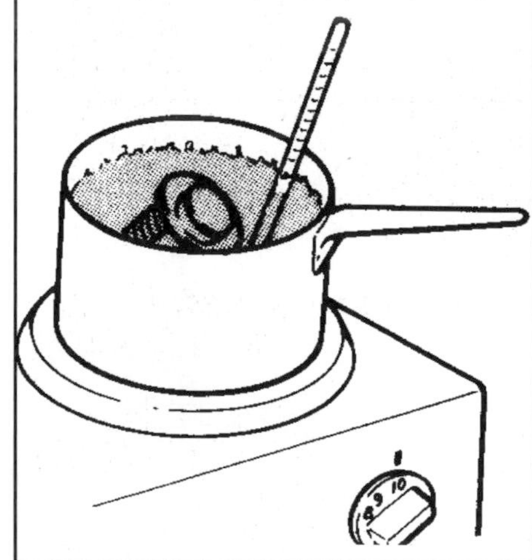

If you've no spare, you'll need to immobilise it in a partially open position to get you home. The engine will probably run too cool, but could overheat in extreme conditions, so watch the temperature gauge if you have one.

ANODES

When two different metals are immersed in saltwater, an electric circuit is produced, acting in the same way as an electric storage battery. A current flows from one metal to the other and one metal is corroded by this electrolytic action.

Raw-water-cooled engines

Seawater flows through the cooling waterways of the engine. Older engines with all castings made of cast iron usually have no anodes, as no dissimilar metals are in contact with the seawater, although some cast iron engines do, in fact, have anodes.

The Petter mini 6 and the mini twin have aluminium castings and do have anodes. Early versions of these engines had ring-shaped anodes incorporated in the water pipes. Many owners were unaware of their presence and disastrous corrosion was possible. Later engines had stud anodes.

Yanmar 1GM and 1GM10 engines have an anode that is hidden from view by the starter motor and is often neglected.

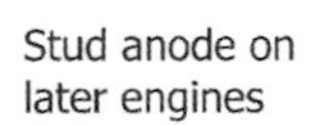

Petter Anodes

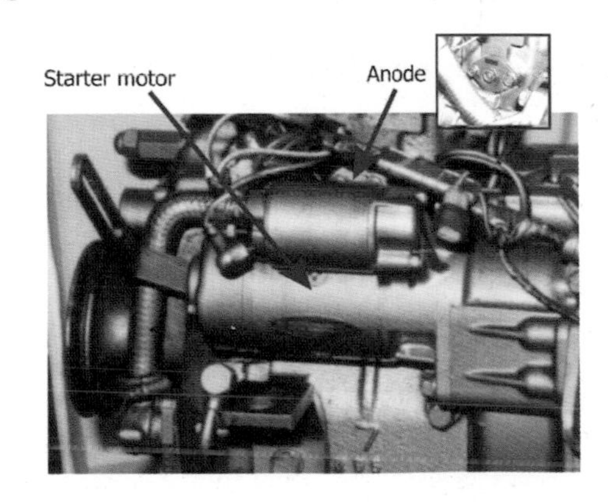

Yanmar 1GM Anode

Freshwater-cooled engines

Freshwater flows through the cooling waterways of the engine. There is no need for an anode in the freshwater part of the cooling system. However, heat exchangers may have dissimilar metals and these may have anodes fitted.

How do you know if there's an anode?

The best place to look is the service schedule, where the fact that the anode needs to be checked or replaced is noted. However, many handbooks don't mention the presence or otherwise of a sacrificial anode, or they note it in the service schedule but give no indication of where its fitted. If different gearbox options are available, you may need to refer to the gearbox handbook. If in doubt, consult your official engine dealer so that you don't miss any anodes.

How often need the anode be checked?

Obviously the engine handbook should be the guide, but it could be prudent to check more often when the engine is new to you, until you have established the pattern for your engine.

Where is the anode fitted?

Any decent handbook should show the location of the anode(s). Some anodes are distinctive – Yanmars have a white label on the head with ZINC printed in red. Volvo Penta anodes have a 'square' head. Others may be anonymous. Where there are multiple heat exchangers, each one may have its own anode.

WARNING – some engines may have different configurations of gearbox, inter-cooler, etc. described in the same handbook, which may cover both direct and indirect cooling systems. Where different heat exchangers – or a different number of heat exchangers – are covered, there may be a different number of anodes fitted. Do check with your engine dealer how many you have, ensuring that he knows the configuration. A misunderstanding here could cost you dear, as a friend of mine found when he had to replace a heat exchanger he didn't know was there.

Tip

Ensure that there's electrical contact between the carrier and the engine block or heat exchanger. This can be checked using a multimeter selected to resistance and checking that there's zero resistance. Without an electrical contact, the anode cannot work, so don't use sealing compound or PTFE tape indiscriminately.

Changing an anode

Engine anodes are generally screwed into an anode carrier. If there's less than half the old anode remaining, replace it with a new one.

1. Unscrew the old anode.
2. Screw the new one into place.
3. Ensure the treads on the carrier are clean.
4. Screw the carrier into place.

WATER STRAINERS

Water-cooled engines use the water in which the boat is floating to cool the engine, either directly or indirectly. This water is likely to contain debris, weed and other contaminants which could block the cooling system. I have been told of a baby eel that found its way as far as the raw water pump!

All engines should have a water strainer fitted in the direct part of the cooling system. Many older boats never had one fitted, and even now some builders fail to fit one. These days, this is usually confined to boats fitted with Volvo Penta sail drive engines, which are supplied with the water intake already connected to the seawater pump with no strainer fitted. The builder may then install the engine as supplied. If your boat has no water strainer, fit one.

Different types of strainer

There are two different types of strainer:

- Those fitted directly to the cooling sea water cock – these may be very small and fairly ineffective, or tall and large, as often fitted to motor cruisers.
- Those fitted to the pipe-work between the seacock and the raw water pump – these are large, can be fitted in a convenient position and some have a transparent top so that they can be checked without removing the 'lid'.

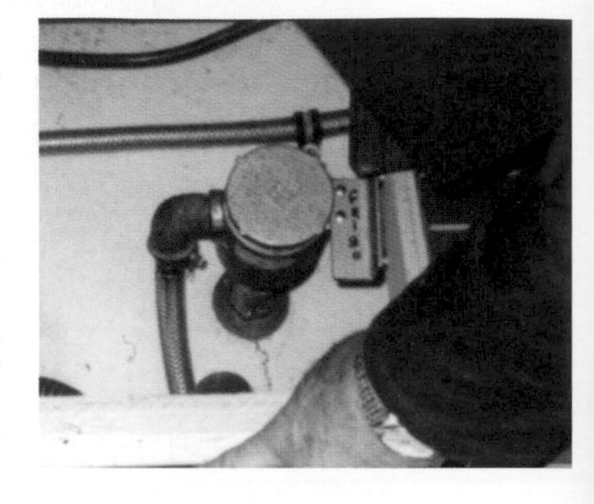

Where is it fitted?

The strainer should be fitted between the raw water intake and the engine's raw water pump in an easily accessible position, as it should form part of your daily check routine.

How often should it be checked?

This depends on the cleanliness of the water. If the water is crystal clear, then weekly will do. If you see rafts of floating weed or small surface debris, then it will need to be checked daily.

Cleaning the filter

If the filter mesh is dirty, clean it in water with a small stiff brush.

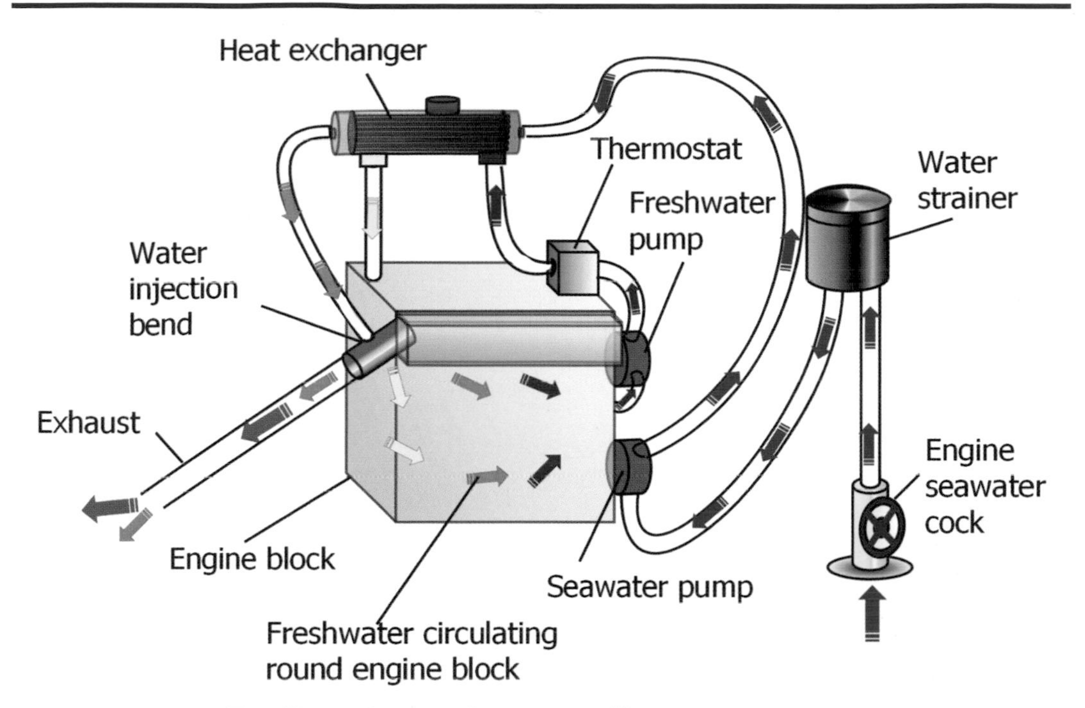

Indirect engine-cooling system

Clearing a blockage

If the strainer is fitted directly to the cooling seawater cock and its top is above the waterline, the cap may be removed, the seacock opened and a rod (piece of broomstick) used to clear any blockage.

For other installations, a pipe joint situated above the waterline and between the strainer and the seacock can be disconnected and a dinghy pump or compressed-air fog horn used to blow the blockage out.

Sometimes a pump will not self-prime properly after the strainer has been checked. In this case, close the seacock, slowly fill the strainer with water and replace the cap. Opening the seacock should then ensure that the pipe is full of water, allowing the pump to prime.

> **Tip**
>
> Ensure that the strainer's cap is replaced properly or air can enter the system, causing the pump to pump air rather than water.

WATER INJECTION BEND

Most leisure boat engine installations have a water-cooled exhaust system. The raw cooling water, on leaving the heat exchanger or engine cooling waterways, is injected into the exhaust system as it leaves the engine.

The injection bend has a hard life. It suffers a mixture of very hot corrosive gas and warm water, which, in the case of a boat used on the sea, is corrosive as well. The exhaust bend should be checked carefully every year for signs of cracking and should be replaced as necessary.

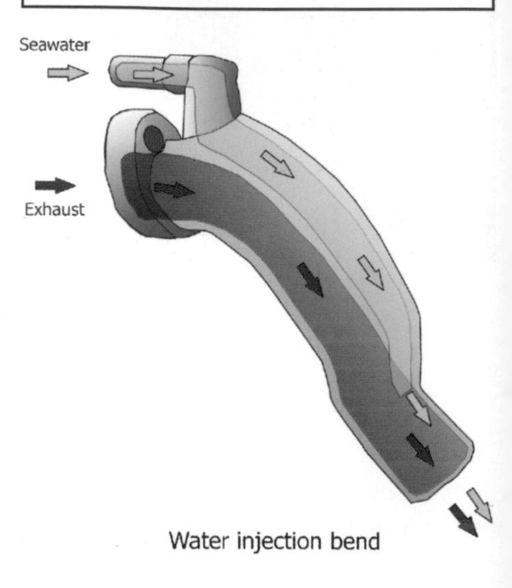

Water injection bend

Sometimes the exhaust bend will last only a couple of years, but seven or eight years is more typical. Some go on for much longer, so it isn't possible to generalise on life expectancy. Checking is the only answer.

Failure may occur anywhere on a cast injection bend but it's more likely at the welds on a fabricated unit. A fabricated stainless steel bend does not necessarily have a longer life expectancy than a cast unit. Where a casting is no longer available, you will be able to have one fabricated by a specialist.

DRY RISER

Where the exhaust pipe exits the engine close to the waterline, it may be advisable to fit a dry riser to help prevent seawater entering the engine via an exhaust valve. This is often neglected by boat builders.

ANTIFREEZE

The freshwater system must be filled with either a mixture of freshwater and antifreeze or a mixture of freshwater and corrosion inhibiter.

Antifreeze will protect the cooling system in two ways: by lowering the temperature at which the coolant freezes and by reducing the amount of corrosion that occurs within the freshwater cooling system.

Protection against freezing

The more antifreeze you use, the lower the freezing temperature, down to a mixture of around 50% antifreeze, when no further protection is achieved.

Protection against corrosion

Similarly, the greater the concentration of antifreeze, the greater the protection against corrosion.

What type of antifreeze should be used?

Reference must be made to the engine handbook, which will tell you if there are any special requirements. Only antifreeze suitable for use in aluminium engines may be used if aluminium is used anywhere in the system. If no specific reference is made, then ensure you follow the instructions on the antifreeze container.

Not all antifreezes are equal. Some have superior corrosion resistance and so, generally speaking, use only well-known brands, especially if you are going to change the coolant every two years rather than annually.

Where freezing temperatures are not expected, a corrosion inhibitor without antifreeze may be used.

How often should the antifreeze be changed?

Reference should be made to the servicing schedule in the engine handbook. Antifreeze performance doesn't reduce significantly with age, so protection against freezing won't deteriorate. However, protection against corrosion will deteriorate with age, so it must be changed regularly. Not only will corrosion occur within the cooling system, but corrosion of the cylinder-head gasket will also occur, causing its premature failure. Testing the coolant for the effectiveness of the antifreeze will give no indication of the condition of the corrosion inhibitor.

If it is necessary to top up the cooling system due to a loss of coolant, don't top up with water only. You must use the correct antifreeze–water mix or the antifreeze will become diluted.

WATER-LOGGED ENGINE

It's possible to fill a sailing yacht engine's working parts with raw water due either to poor installation or careless operation.

On a sailing yacht, the exhaust pipe is normally swept upwards above the waterline to prevent following seas flooding the exhaust. This requires that the engine must be running, so that the exhaust gases can expel the cooling water 'uphill' and out of the exhaust pipe.

In modern yachts with a shallow canoe body, the exhaust has little room to fall from the exhaust pipe to the lowest point of the exhaust system. It then becomes even easier for water to back up into the exhaust manifold. A dry riser, fitted to the exit from the exhaust manifold, increases the fall from the engine to the water trap (see the photograph earlier in this chapter).

In order that raw water can't siphon past the raw water pump and into the exhaust pipe, the water injection pipe must be placed at least 150 mm above the highest

Tips

- To reduce the build up of limescale in the system, use distilled or deionised water rather than tap water to mix with the antifreeze. If you get your de-ionised water from a car accessory shop, it isn't expensive.
- To increase the engine efficiency, some engines are designed to run at a temperature exceeding 100 °C. Their thermostat is correspondingly set higher, and the pressurisation of the hot water in the system raises the water's boiling point to above the high running temperature. The boiling point of the water depends not only on the pressure setting of the filler cap, but also on the fact that a 50% mixture of antifreeze is used. If no antifreeze is used, the coolant will boil before the high temperature alarm is activated, so you will have no indication if a serious overheating fault arises. The Beta engine handbook warns of this eventuality.

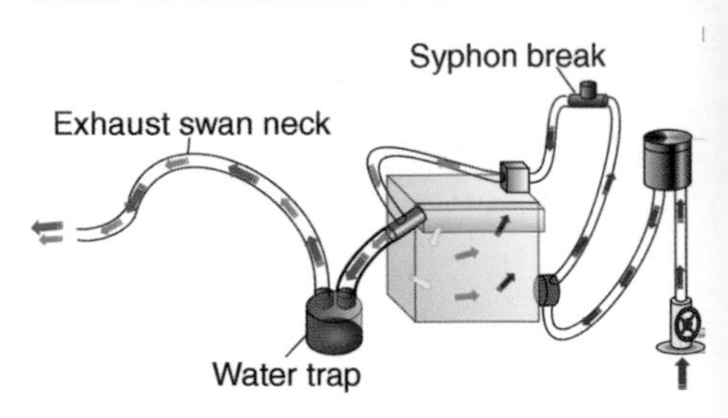

Sailboat direct engine-cooling system

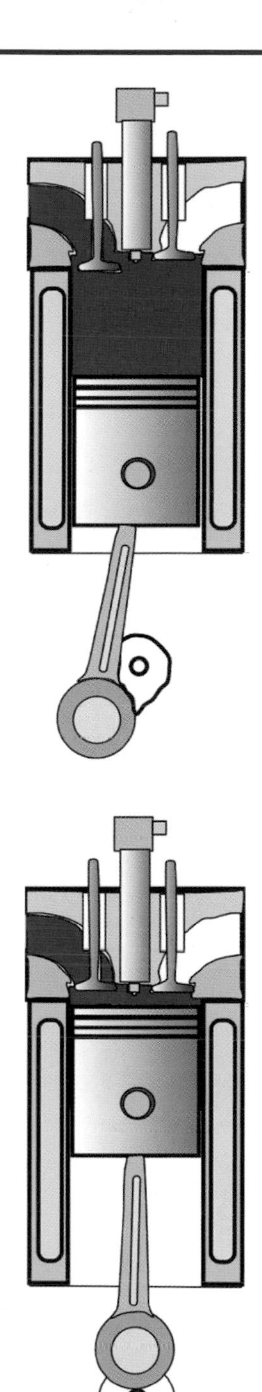

heeled waterline. If this can't be achieved, it's essential to fit a siphon break in the cooling system and to ensure that if this is of the anti-siphon valve type, it receives annual maintenance.

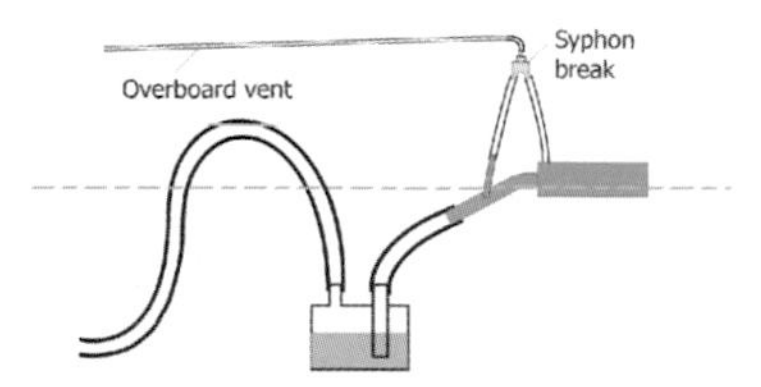

Syphon break in cooling system

To ensure that raw water cannot rise to a dangerous level in the exhaust, it's essential that the engine is not cranked for too long a time in an attempt to start it. A good rule of thumb is to use no more than two 15-second attempts to start the engine. If the engine won't start after this time, the cooling seawater cock must be closed until the engine starts. Ideally, you should investigate why the engine won't start in a shorter period of time.

If water gets into the cylinder, because the compression ratio is so high, there may not be enough room for it with the piston at the top of the compression stroke. The result of this may be a bent connecting rod or, in an extreme case, a shattered piston.

Only a small bend in the connecting rod can result in a lowered compression ratio sufficient to prevent proper running, or even to prevent the engine starting from cold.

Water entering the working parts of the engine can cause serious corrosion, which will require the engine to be rebuilt at great expense.

The engine shown in the photograph suffered fatal damage because the engine had been cranked continually because it wouldn't start, due to running out of fuel. The engine was run in this condition for two weeks before the problem was noticed. Had the oil dip-stick been checked daily, as required, this problem would have been noticed the following day, and the cost of a new engine may have been avoided.

DRAINING AND REFILLING THE COOLING SYSTEM

You will need to drain the cooling system to change the antifreeze of a direct cooling system, and sometimes you may need to drain the raw water system.

Most engines have drain taps located at several points in the engine block and often the water-cooled exhaust manifold as well. The drain ports are situated at the lowest points of each casting and there may be

Water drain (Yanmar)

several such 'lowest points'. Unless these drain taps are opened on a regular basis, they are likely to become seized or blocked, especially on a raw water-cooled engine.

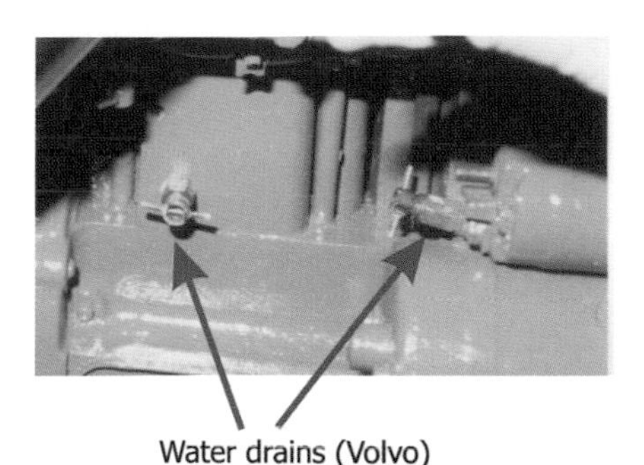

Water drains (Volvo)

Water drain (Perkins)

Bleed points

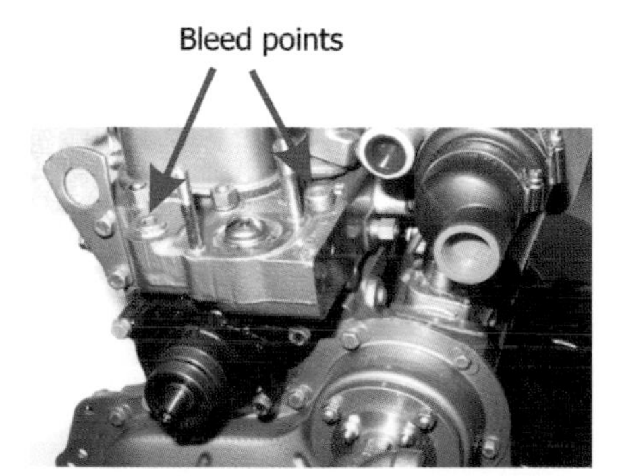

Perkins 4108 bleed points

Some engines have bleed points that need to be opened to allow air to escape as you fill the system. Do follow any special instructions contained in the handbook to ensure the system is filled properly.

Unless you can arrange to catch the water, it will run into the bilge from where it will need to be pumped out.

Some engines pass the cooling water through a *calorifier* to provide domestic hot water. The heating coil in the calorifier and its supply and return pipes become part of the cooling system and will add to its volume, which must be taken into consideration when calculating the volume of antifreeze used.

Raw water systems

Draining

1. Close the cooling water intake seacock.
2. Open the water strainer to allow air into the system.
3. Open the drain valves if possible.
4. Remove the raw water impeller.

5. Some engines pass the raw water through the gearbox casing prior to the raw water pump. In this case, there will be a drain point on the water cooling jacket as well – open this. Don't confuse this drain point with an oil drain plug.

Pockets of water may remain in the system, especially if you are unable to use any of the drain valves. This is why I don't recommend draining a raw water system as a method of frost protection. Not only that, but there will be (salt) water and air in the cooling ways, and this will accelerate the corrosion process (see Chapter 15 on winterisation).

Refilling

1. Remove the thermostat.
2. Close all drain valves.
3. Replace the water pump impeller.
4. Close the drain point on the gearbox cooling jacket if applicable.

EITHER

5. If the water strainer is higher than the thermostat housing, fill the system by filling the strainer body until water comes out of the thermostat housing.

OR

6. If the water strainer is too low, disconnect a pipe from somewhere convenient prior to the water pump, extend this vertically to a level above the thermostat housing and fill or attach a mains water pipe to this point and fill gently.
7. Observe any special bleeding procedures.

The necessity to carry out this procedure can be avoided by flushing the raw water system through, as described in laying up a raw water cooling system.

Freshwater systems

Draining

1. Open the water filler cap to allow air into the system.

2. Open the drain valves. If they won't open, then the system cannot be drained fully.
3. Disconnect a freshwater hose as low down in the system as possible to speed up the draining process.

Refilling

1. Close all drain valves.
2. Reconnect any disconnected pipes.
3. Fill the system slowly through the filler cap. When the system seems to be full, allow time for water to flow into remote parts of the system and continue filling until it will accept no more.
4. Observe any special bleeding procedures.

Lubrication

Lubricating oil is contained in the engine oil *sump* located at the bottom of the engine. This is the equivalent of the oil 'tank'. An engine-driven oil pump sucks oil from the sump via a coarse strainer, immersed in the sump. The oil, under pressure, is delivered to the oil filter, after which it is distributed by means of pipes and oil *galleries* to all the parts of the engine requiring lubrication. The oil returns to the sump by dribbling back down the sides and working parts of the engine. Once the engine is stopped, it takes some time for all the oil to return to the sump.

THE SUMP (OIL PAN)

The sump is the lowest part of the engine. On sailing boats it often needs to be deeper than its standard 'automotive' cousin to ensure that the strainer stays immersed in oil even when the boat is heeled. The engine handbook normally states the maximum heel allowed when the engine is running. Commonly it is between 25 and 30 degrees.

The walls of the sump allow heat to be extracted from the hot oil by air passing around it, so it's essential to allow free circulation of air around the sump.

Unlike a car, there's normally insufficient room below the engine to allow old oil to be drained into a container.

Consequently, old oil must be sucked out of the sump by means of a pump when it's time to change the oil. Some engines have a built-in hand pump to achieve this, but commonly you need to insert the pump's suction pipe into the engine oil 'dip-stick' hole.

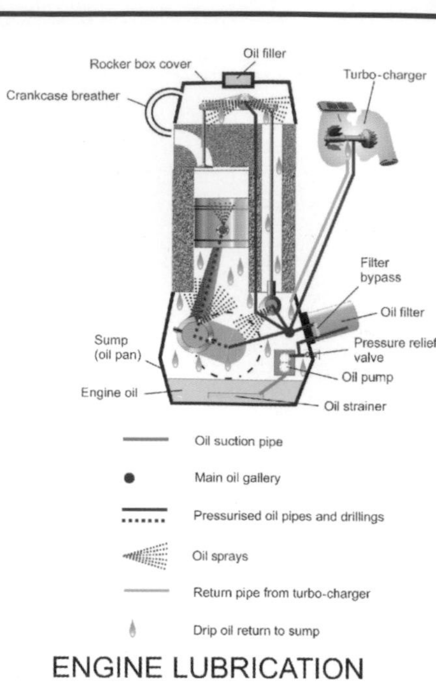

ENGINE LUBRICATION

THE STRAINER

The strainer is in the bottom of the oil sump and prevents solids (there shouldn't be any) from being sucked into the pump. On most modern engines the strainer is accessible only when the sump is removed and is not part of the normal servicing schedule.

THE OIL PUMP

A few engines use a pump with two intermeshed gearwheels to pressurise the oil, but most use a trochoidal pump. Usually, the rotor has four lobes and the free rotor five. The free rotor always has one more lobe than the rotor. The rotor is driven by the engine. The free rotor, in which the rotor sits, is driven by the rotor. The spaces between the lobes

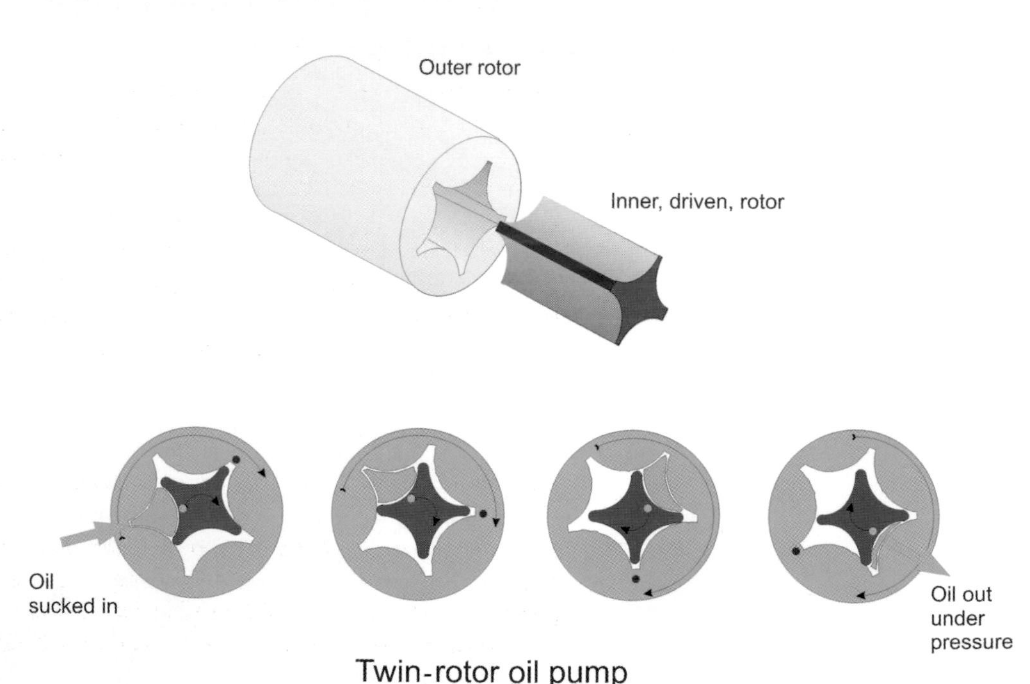

Twin-rotor oil pump

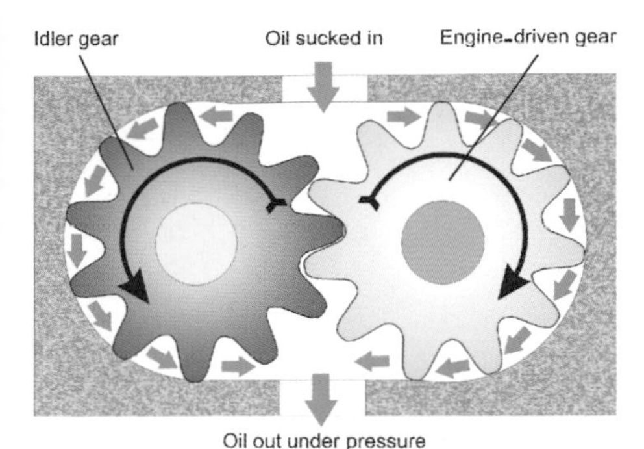

Gear-driven oil pump

vary constantly and as the volume reduces, the oil is pressurised.

More often, the oil pump is of the gear type.

The oil pump is driven by the camshaft. It has a long life and normally needs attention only during major engine overhaul. Full details of servicing the oil pump will be found in the engine's workshop manual.

The oil pressure range from engine idle rpm up to maximum rpm is given in the handbook or workshop manual. Most small engines don't have an oil pressure gauge, so you have no way of checking this other than ensuring the oil pressure warning light extinguishes once the engine is running. On engines with high running hours you may find that the oil pressure lamp flickers at idle rpm when the engine is hot. Provided the light extinguishes as the rpm rises, this should not be a worry, but you can get a competent engineer to check the actual oil pressure by temporarily replacing the warning light sensor with a pressure gauge. You could also increase the idle rpm a trifle if this is of concern.

PRESSURE RELIEF VALVE

If the oil pressure is too high, such as when starting a cold engine in cold weather, damage could be caused to some components. The *pressure relief valve* allows some of the oil to be returned to the sump, to reduce the oil pressure to an acceptable level. On some engines, the bypass relief spring is accessible from outside the engine if adjustment is required at any time. However, adjusting this valve should be treated with caution, as the cause of low oil pressure could be dirt on the valve seat or a worn oil pump.

THE OIL FILTER

The oil filter is removable and needs to be replaced by a new one according to the service schedule. Its purpose is to remove from circulation any solid particles, such as carbon from the combustion process or metal particles due to wear of metal components. Remaining in circulation, these particles would increase the rate of wear within the engine.

Decent oil filters will have a bypass so that unfiltered oil will still circulate should the filter become blocked. They will also have a non-return valve to prevent oil in the pressure lines from draining back into the sump when the engine is not running, to ensure that oil is present at the working parts as the engine is started. Cheap replacements may not have these features and may also not have such a fine or efficient filter. Weighing a cheap filter in the hand may well reveal a considerable difference from the real thing.

The oil filter needs to be changed as scheduled by the handbook, but where the engine gets little use or never reaches full operating temperature, more frequent changes may be appropriate.

CRANKCASE BREATHER

As the piston(s) move(s) up and down, pressure pulses occur in the crankcase. Additionally, any combustion gases blowing past the piston will increase this pressure, which, if not released, can cause loss of oil. The excess pressure is felt in the rocker box and it's normal for the gases to be fed into the air intake, rather than directly to the atmosphere, as the gas is likely to carry oil fumes with it. These fumes are then ingested by the engine and burned together with the normal fuel. If there are a lot of oil fumes in the breather gas, it's possible for the engine to continue running using this as fuel, even when the fuel is shut off by the stop control. It's even possible for the engine to overspeed, as these fumes are not controlled by the engine's governor. The engine may not respond to the stop control or even to shutting off the fuel at the tank. In this case, the only way of stopping the engine is to block the engine air intake with a solid object such as a block of wood.

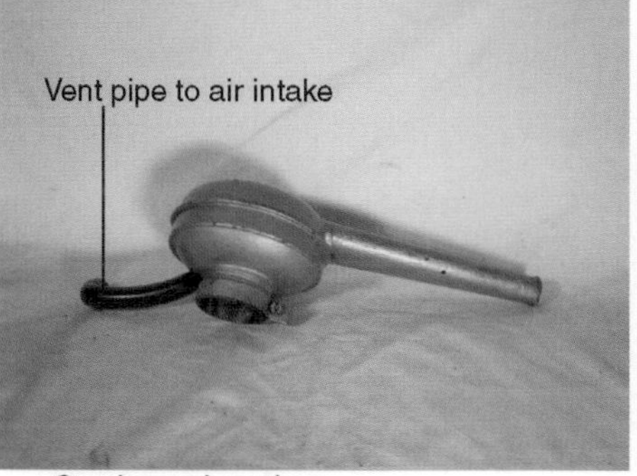

Crankcase breather

The breather system on most engines is very obvious and the handbook may call for periodic cleaning.

Volvo Penta 2000 series engines have a hole in the rocker box gasket, which aligns with a hole in the cylinder

head, so it's essential to make sure the gasket is fitted the correct way up.

ENGINE LUBRICATING OIL

Lubricating oil fulfils several functions:

- it lubricates the working parts of the engine to reduce friction and wear;
- it cools parts of the engine that can't be reached by the water-cooling system;
- additives in the oil reduce the effects of corrosion by the byproducts of combustion.

Lubrication

The oil needs to be 'sticky' enough to cling to the parts requiring lubrication when the engine is not running, so that when it starts, there is already some protection before the circulating oil reaches that spot. This requirement is the opposite to that requiring the oil to flow easily so that it reaches its destination quickly.

The oil needs to be 'thin' when it's cold, so that it will flow easily and not create too high a pressure, but 'thick' when it's hot, so that it will still flow easily but not allow the pressure to fall to too low a value. This is achieved by using a *multi-grade* oil, indicated by having two viscosities, such as 10W40, indicating that it acts as if it were 10 grade when cold but 40 grade when hot.

The oil needs to resist being squeezed out of where it's needed by high-contact forces between two parts, and this is where the additives play a big part.

From the oil filter, the oil passes to the oil gallery from where it is distributed to those parts requiring it. Some will flow under pressure by pipes situated within and outside the engine casing. Some will pass though oilways drilled in various components to reach their target. Oil sprays are used to lubricate and cool parts inaccessible to the former, such as pistons, cylinder walls and valve gear. Oil pressure is also used to operate timing chain tensioners where fitted.

Cooling

Where components need to be cooled, oil is sprayed onto surfaces, such as the inside of the hollow piston, to take away the heat of combustion. As combustion temperatures can be close to 1000 °C, the inside of the piston gets very hot and so does the cylinder wall, so the oil must not lose its lubricating properties under these conditions. The conditions are even more severe when it comes to cooling a turbo-charger.

Corrosion

One of the byproducts of combustion is sulphurous acid. At normal engine running temperatures this is volatile and most of it will 'boil off' before contaminating the lubricating oil. If the sulphurous acid doesn't reach sufficient temperature, or cools, it becomes sulphuric acid and will then contaminate the oil, leading to significant corrosion within the engine.

Good diesel engine lubricating oil will have corrosion inhibitors incorporated. These inhibitors will be used up in doing their job, and the cooler the engine is run, the more quickly they will be exhausted. Light, or infrequent, engine use will require the oil to be changed more frequently than specified in the handbook.

The oil specification

Diesel engine lubricating oil is normally listed under the API (American Petroleum Industry) system in the form of API xx-yWzz.

- xx indicates the specification, especially of the additives, according to the type of engine and use. Grade CC is for normally aspirated, light-duty engines and CF is for hard-worked turbo-charged engines. It doesn't benefit the engine to use a higher grade than specified.

- yWzz indicates the viscosity of a multi-grade oil, such as 10W40. This is chosen according to the ambient temperature of the air where the engine will be used. There is often a temperature table in the engine handbook indicating the viscosity you should use.

Some oils are 'synthetic', which, in automotive use, can be beneficial, if expensive. New engines are normally filled with 'running-in' oil and this needs to be changed at around 50 hours' running. The formulation of synthetic oil reduces friction but doesn't aid running-in, in fact it hinders it. As most marine diesel engines in leisure use rarely become fully run-in, synthetic oil isn't recommended by most marine engine manufacturers. When challenged by the legal department of a major petroleum company about this statement given in answer to a PBO reader's query, it all went very quiet when I asked for confirmation that no harm would result from its use in the leisure environment. You can draw your own conclusion.

What lubricating oil should you use?

The engine handbook will specify the type of oil needed. If the engine is not a current model, the oil specified may not be obtainable. In this case use the closest – if a 'CC' oil was specified but only CD or above is available, use CD. Older engines may specify single-grade viscosity, such as 30 grade. No harm will result from using a multi-grade; in this case, a 10W30 will be fine, though you may get a flickering oil warning light at idle rpm when the engine is hot (see above).

How often should the oil be changed?

Again, the handbook will specify oil change intervals, which should not be exceeded. An annual oil change should be the minimum, even if the engine hours achieved have been low. The best time for an annual oil change is just before you lay the boat up for the winter, as then oil with fresh anti-corrosion additives will be in place for the period the boat is out of use.

Changing the oil

Engine oil change

1. Most people recommend that the engine should be run to warm up the oil, making its extraction easier.

2. Then let the engine stand for 10 minutes to allow the oil to settle.

3. Remove the dip-stick and insert the tube of the oil extraction pump, trying to get it as close to the bottom as possible.

4. Pump out the oil.

5. Pour the required quantity of new oil into the oil filler. (This won't be contaminated by the dirty oil still contained in the old filter, but will give time for the oil to reach the sump).

6. Using a suitable wrench, remove the old oil filter, trying to contain the spilled oil.
7. Fit a new filter, first lubricating the oil seal, tightening as indicated on the instructions printed on the filter.

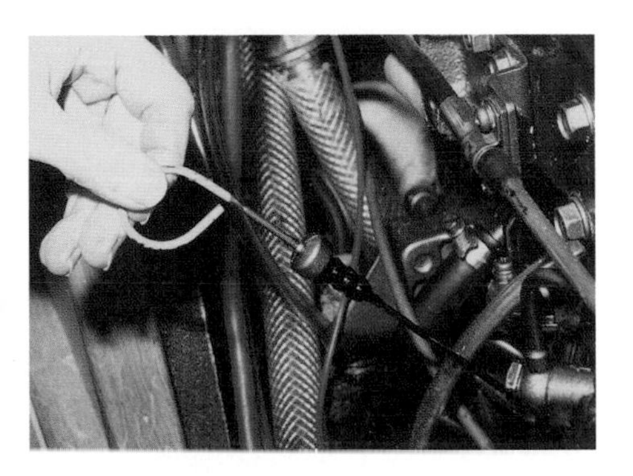

8. Check the oil level. (As the filter doesn't yet contain any oil it may over-read).
9. Run the engine for a couple of minutes to check for leaks. If you have a mechanical 'stop' control, keep it pulled until the oil pressure light goes out, to allow oil pressure to build before the engine starts.
10. Wait 10 minutes, check the oil level and top up if necessary.

Gearbox oil change

1. Oil normally has to be removed by a pump, through the dip-stick hole.

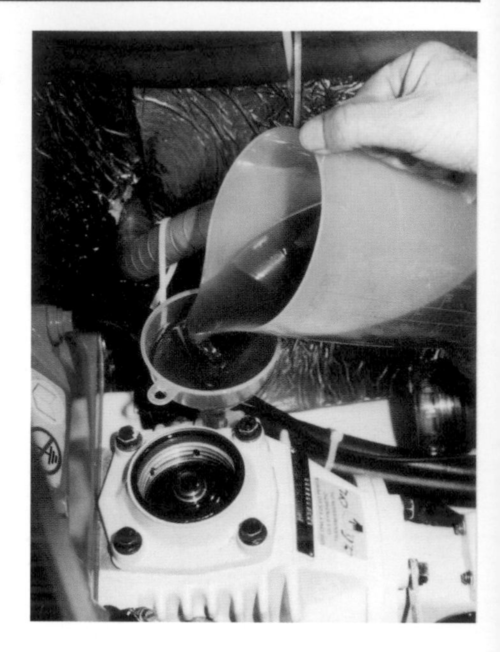

2. Refill with oil as specified – it may be the same specification as used in the engine or it may be completely different – sometimes the gearbox will have its own separate handbook.

3. For sail drive legs and outdrives, the boat will have to be out of the water to change the oil, as oil is drained from the bottom of the leg. However, some modern sail drive legs can have their oil changed whilst still in the water by pressurising the leg. You will need to look at your handbook to see how it's done on your specific leg.

The Air System

The air system is deceptively simple, but any defects can have large repercussions. It comprises:

- the air entering the engine compartment;
- the air supply via the air intake;
- the exhaust system.

In effect the diesel engine is a large and efficient air pump and is capable of a large suction at the air intake. A 30 hp diesel engine running at 3000 rpm will consume around 1000 litres of air per minute, even when it is developing little power. That's a box of air 1 metre × 1 metre × 1 metre every minute.

Many boat builders install a 10 cm diameter pipe to supply air to the engine compartment, and the engine will suck through this pipe air at a velocity of around 4 metres per second without any help from the small fan that some boat builders seem to fit. Other boat builders supply no dedicated air supply with no detrimental

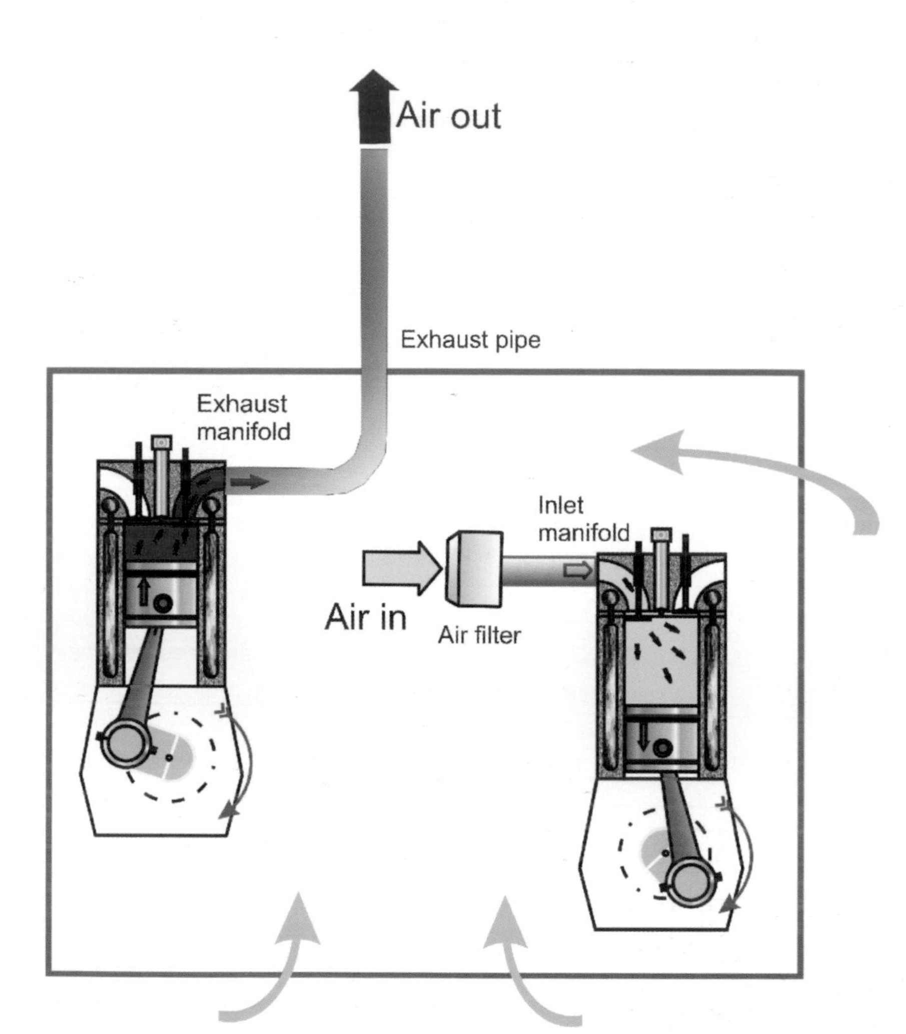

The air system normally aspirated

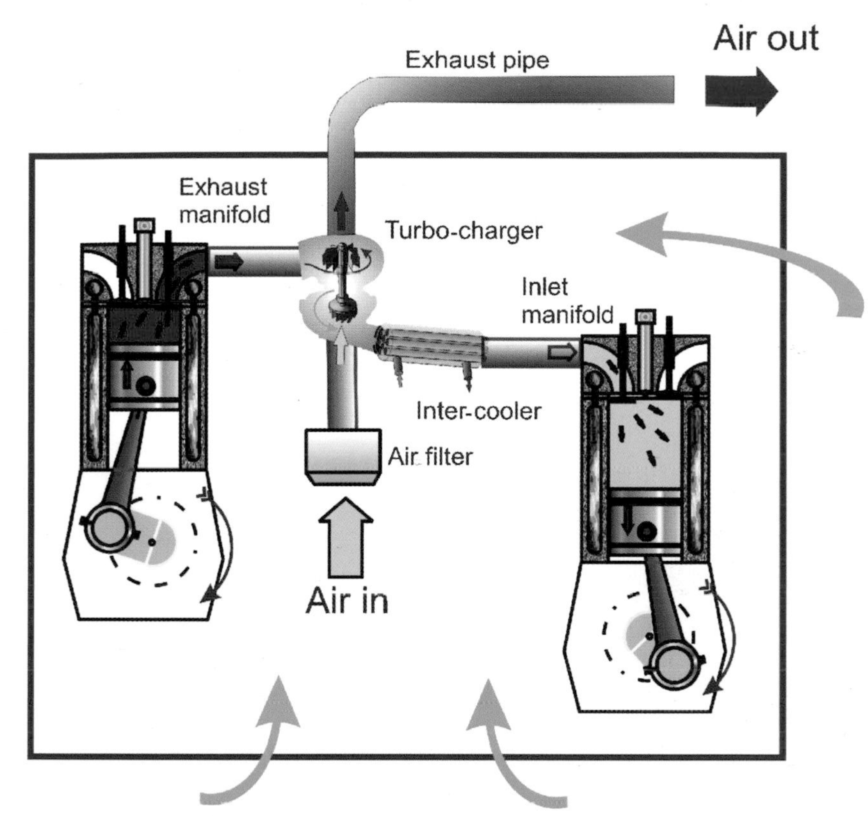

The air system turbo-charged

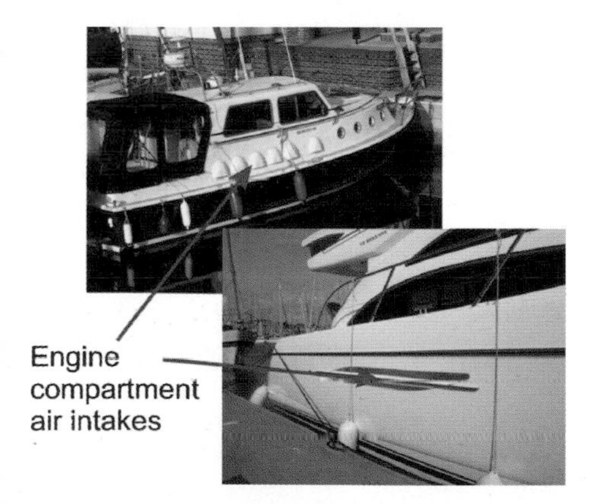

effect. Engines of high-powered motor cruisers need large air intakes.

Before the air enters the engine to support combustion, it is used to cool the engine casing. Something like 25% of the cooling requirement is met in this way. It's important that the air supplied is able to flow around the engine casing prior to being consumed, so any dedicated air supply should not be directed at the engine air intake. Rather, it should be directed at the oil sump. Where there's no dedicated air supply, the air is sucked in from many directions and cools the engine just as efficiently, provided that the engine is relatively small.

If the air supply is insufficient to support combustion, black smoke will be seen in the exhaust and this will cease if an engine access hatch is opened. Should this occur, then a better air supply must be arranged.

Air supply can be diminished if the exhaust is partly blocked, because if there's no room to let the air out, it can't get in. Partial blockage can occur due to a build up of soot or to delaminating and softening of the pipe.

THE AIR FILTER/SILENCER

Normally an air filter is fitted to the air intake. The filter element may be of sponge rubber, paper or wire wool. Sometimes there's only an empty box acting as a silencer, as generally the air within a boat's engine compartment is not laden with dust and dirt, as it may be on land.

Maintenance should be carried out according to the engine service schedule.

Sponge rubber filters

These are washed in water to which a gentle detergent has been added, rinsed and hung to dry before being replaced.

Paper filters

These are contained in the filter box and replaced by new ones as required.

Tip

If there's a length of rubber pipe between the air intake and the engine, any softening or delamination of this pipe can cause a reduction of airflow as suction increases. If delamination is the cause, this may not be visible until you remove the rubber pipe.

Wire wool filters

The filter is washed with paraffin, by immersion and swilling around. The element is then oiled by dribbling light engine oil onto it, and is then rotated to distribute the oil. The oiled wire catches any dust and dirt.

THE EXHAUST

Generally this gets little or no attention, despite its importance.

- Most deterioration will occur at the exhaust bend, described in the chapter on cooling. This must be inspected annually for any cracks or leaks and replaced as necessary. Keeping the external surfaces clean will aid this process.

- Annual disconnection of the exhaust hose from the injection bend will allow both to be inspected for carbon build-up and for any delamination of the hose.

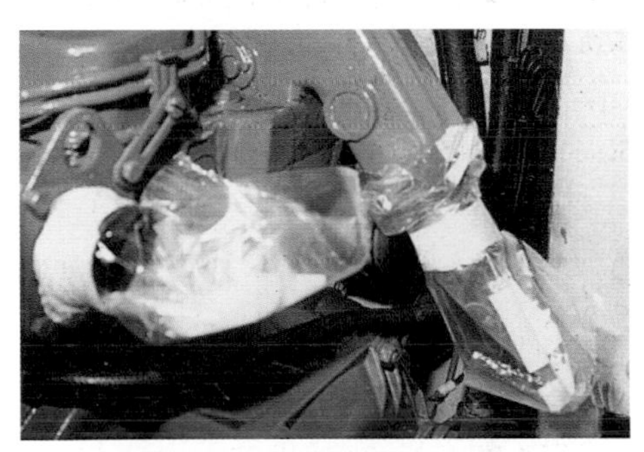

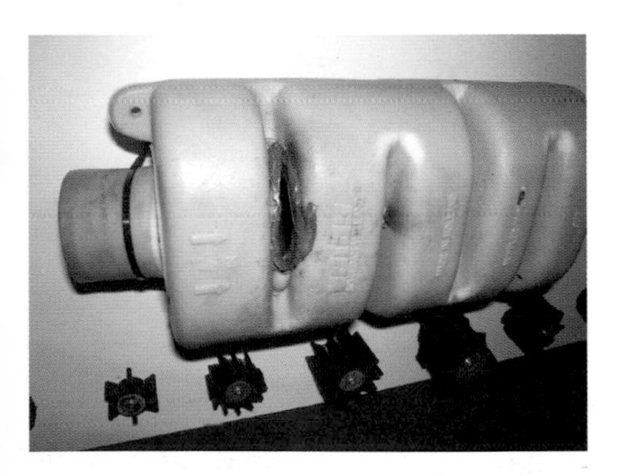

- Annually inspect the water trap and silencer for damage, especially for corrosion of the welds of a metal one.

- Check for deterioration of the fitting where the exhaust leaves the hull.

Engine Electrics

If we remember that a basic marine diesel engine requires no electricity to run, then the electrical system to support the engine should be very simple. A simple 'hand start' diesel needs no electricity at all.

Few boats today have no ancillary electrical components, such as cabin and navigation lights, electronic navigation equipment, etc., so the basic engine electrical system takes these requirements partially into account. However, it's unlikely to support extended cruising or the fitting of 'lifestyle' equipment, such as a TV, microwave oven, etc.

Basically, the engine electrical system is straightforward and easy to understand if you have a block diagram, which you can use in conjunction with the engine's normal wiring diagram.

TYPICAL ENGINE ELECTRICS BLOCK DIAGRAMS

The start circuit diagram allows you to visualise how the parts of the system are joined electrically. You can use the engine block diagram in conjunction with your own engine wiring diagram to get the colour of the wires. If you copy this diagram and cross out the bits

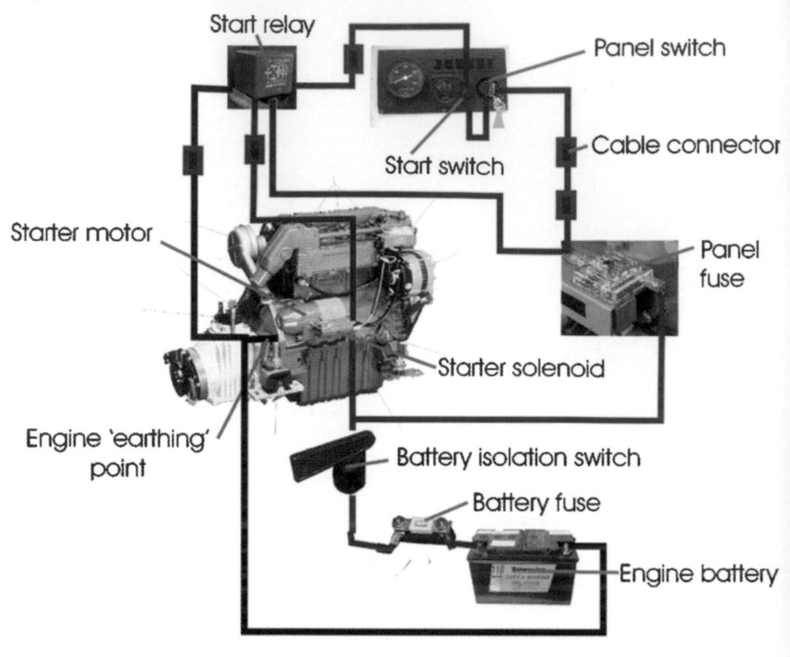

Only the wiring associated directly with the engine start circuit is shown

that don't apply to your engine, you are halfway along the road to troubleshooting.

The battery

The battery supplies power to start the engine.

The battery switch

The battery switch is used to isolate all the electrical circuits from the battery.

The alternator

The alternator generates electricity to charge the battery and to supply electrical power to those items requiring it. On older engines a dynamo would have done this job, but less efficiently. The alternator is belt-driven by the engine.

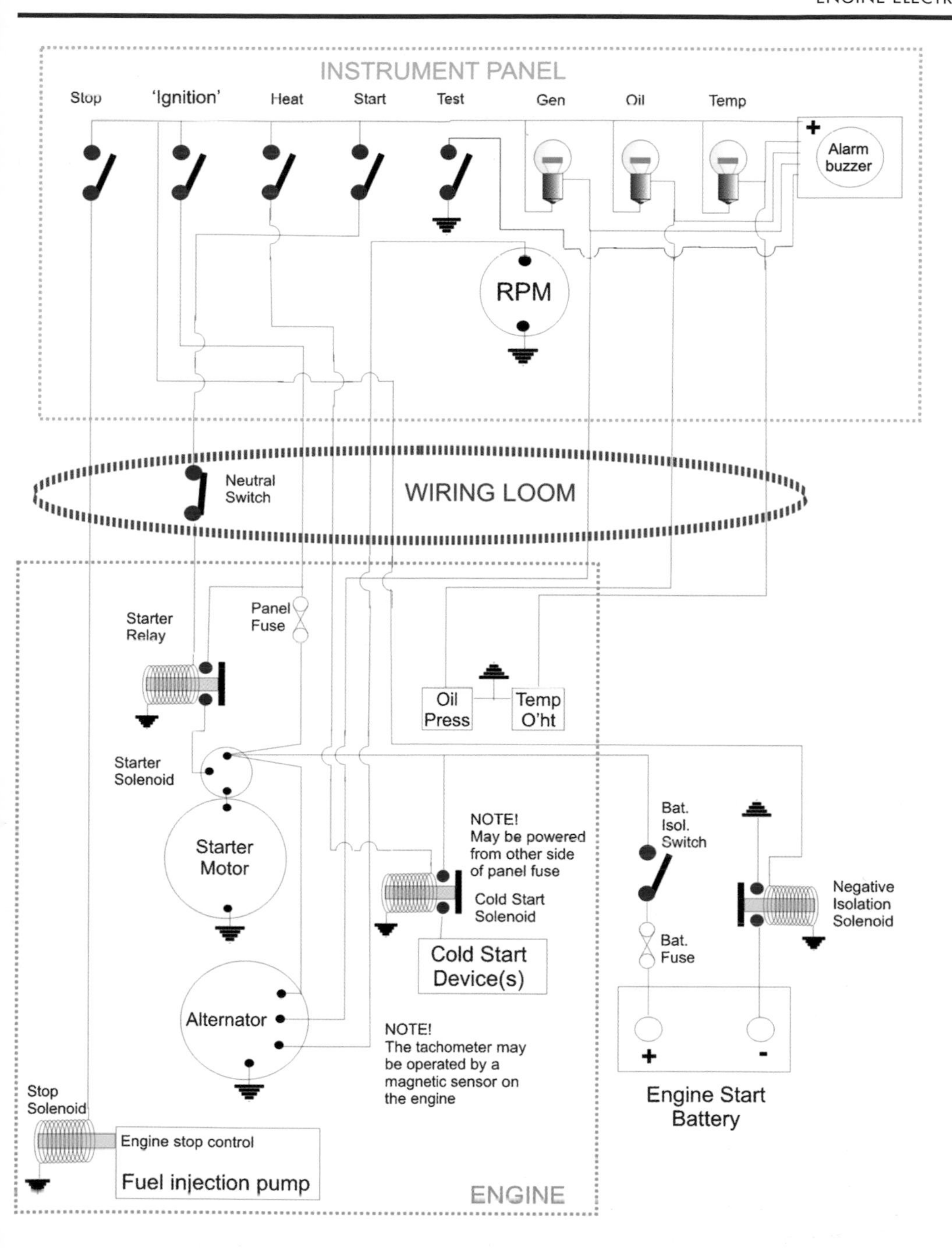

ENGINE ELECTRICS BLOCK DIAGRAM

The starter motor

The starter motor is an electric motor powered from the battery and is used to start the engine.

The starter solenoid

The current required to start the engine is very high. If this current had to pass through the starter switch, the cabling and the switch itself would need to be very heavy duty, and hence expensive and unwieldy. The solenoid allows a low-current circuit to operate the solenoid's electromagnetic coil and contacts, which, in turn, then allow the high starting current to flow. On modern engines, the solenoid also activates the starter pinion clutch.

The starter relay

If the circuit between the starter switch and the starter solenoid is too long, high resistance may make the starter operation unreliable. In this case an intermediate electromagnetic relay may be used to boost the current to the starter solenoid.

The cold start relay

If cold start devices, such as heater plugs, are used, these will be activated using a relay to handle their operating current. This is usually operated by holding the start key at the 'heat' position for the time specified before continuing with the engine start, or by pressing a 'heat' button. On engines with full electronic engine management systems, this is activated automatically as required.

The stop solenoid

On some engines the lever used to cut off the fuel supply at the fuel injection pump to stop the engine is operated by an electromagnetic *stop solenoid* rather than by pulling a stop handle. It is operated by holding the start key at the 'stop' position or by using the 'stop' button until the engine stops. In an automotive application the stop solenoid is powered to run. This means that if you lose electrical power, the engine will stop. In most marine applications the stop solenoid is powered to stop and requires no power for the engine to continue running. Engines used to power a generator (gen. set.) are run unsupervised, so any fault detected is required

Tips

- The negative isolation solenoid will not appear on the engine electrical wiring diagram as it's part of the 'boat' and the owner may be completely unaware of its presence. In the case of the engine failing to start, this may be the culprit.
- Solenoids which spend a long time energised have two sets of contacts to supply the power. The first set operates the action of the solenoid, which takes quite a bit of current, and once the contact is made, another set of contacts is energised to 'hold' the solenoid engaged, which takes far less current. The operating current is then switched off automatically. When setting up or troubleshooting this type of solenoid, it's necessary to check both sets of contacts for correct operation.

to shut down the engine automatically. In this case the stop solenoid is powered to run and any fault detected just switches off the power and stops the engine.

The 'negative' isolation solenoid

Steel and aluminium hulls can suffer from electrolytic corrosion. Ideally, the engine electrics should use a two-wire system, where all negative wires are actual cables, rather than using the engine block itself as the negative 'wire'. This requires special and expensive items such as the starter motor, and so is not always used. For economic reasons these installations often use normal single-wire components. To help minimise corrosion where normal single-wire components are used, a relay is sometimes introduced into the negative cable between the battery negative terminal and the engine block. This solenoid is activated by the engine start switch so that the negative circuit is activated only when the engine is running.

The 'ignition' switch

As a diesel has no 'ignition' circuit, the term 'ignition switch' is a misnomer, but it's understood by all. This switch is often multifunctional and in its 'RUN' position, the instrument panel is energised. Other positions may include 'HEAT', 'START' and 'STOP', or there may be separate switches for these functions.

The engine 'fuse'

Most engines have a fuse or circuit-breaker in the supply (+ve) wire to the engine panel. This protects all the engine circuits except the starter 'power' circuit. This fuse isn't always mentioned in the handbook (Yanmar), though its failure will render the engine 'dead'.

The battery fuse

This is rarely fitted by boat builders. Its omission could lead to an electrical fire if the supply from the battery to the engine suffers a short circuit. An argument for not fitting one is that its failure with the engine running will result in the alternator's diodes being destroyed. Personally, I would prefer that to the boat being destroyed by fire!

The tachometer

This instrument indicates the engine rpm. It may be driven in one of two ways:

- From special connections on the back of the alternator, in which case belt slip or a change of alternator pulley size will give incorrect readings. Generally it doesn't matter what make of tachometer you use, as long as it's set up for the pulley ratio.

- From a dedicated transducer situated within the engine. These generally count magnetic pulses and you need the correct proprietary tacho.

Although a tachometer is a very good troubleshooting aid, one is often not fitted because of the extra cost.

Warning lights

Sensors on the engine detect low oil pressure, high coolant temperature and no output from the alternator (dynamo). These cause their respective warning lights to illuminate when a fault occurs. If you're lucky, there may also be an indication of cooling water flow.

When the 'ignition switch' is turned 'ON':

- The oil low pressure warning light should illuminate as there's no oil pressure.

- As there's no alternator output, its warning light should illuminate as well.

- As the coolant temperature is not too high, this warning light will not illuminate, so there should be a 'TEST' switch to check that this light is working.

- Yanmar sail drive engines also have a warning light to indicate that there's water between the internal and external sail drive leg-sealing diaphragms. This should not illuminate when the 'ignition' is turned 'ON', so the 'TEST' switch will test this as well.

- If cold start heater plugs are installed, an indicator light will illuminate when you select 'HEAT'. This is not tested with the 'TEST' switch.

> **Tip**
>
> Volvo Penta uses a circuit that prevents the warning sounding when you stop the engine, so there's no reminder to switch the 'ignition' 'OFF'.

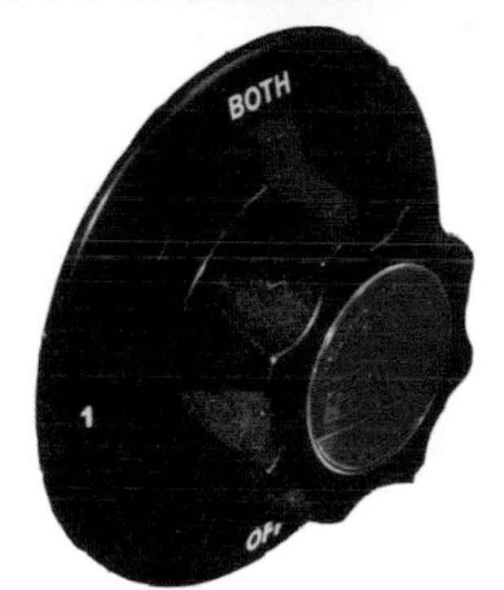

Warning 'buzzer'

- As an 'attention getter', a warning buzzer will sound when any warning light illuminates.
- This warning buzzer should sound when you switch the 'ignition' 'ON'.
- Generally, when you stop the engine the alternator warning light will illuminate, as there's no longer any output, and within a short period of time the oil pressure warning light will also illuminate.
- The buzzer will sound until you switch the 'ignition' 'OFF'. This is a good reminder to switch the 'ignition' 'OFF' when you stop the engine.
- If the engine is stopped by switching and holding the 'ignition' key at 'STOP', the buzzer will stop as soon as you release the key.

ENGINE-DRIVEN 12-VOLT DC ALTERNATOR

Depending on the size of the engine, the alternator will usually have a maximum output of between 35 and 60 amps as standard, although it's possible to fit one of higher output. Higher outputs often require a multi-belt drive to prevent belt slip.

Modern alternators (as opposed to DC dynamos) generate alternating current (AC) and convert this into direct current (DC) to charge the battery. The voltage is regulated at a nominal 12 volts (or 24 volts) to charge the battery and supply the electrical loads.

The battery must *never* be disconnected from the alternator when the engine is running, or the alternator's 'diodes' will be destroyed. In other words, DO NOT switch off the engine battery switch when the engine is running. Not all battery switches are of the same quality, and unless yours is marked with a well-known maker's logo, one circuit may be broken before the next has been selected. If in doubt, don't even change the battery selection with the engine running. Some 'single' battery switches may lose contact, as they wear even when switched on.

Alternator output

You will be wrong if you think that a 60 amp alternator will deliver 60 amps as a matter of course. Three things materially affect output: alternator speed, alternator temperature and the load to which it's connected.

Alternator output will depend on how fast it is turning (rpm). Any particular alternator has a maximum speed, and so its driving pulley size is calculated so that this maximum rpm is not exceeded at the engine's maximum speed. For instance, if the alternator maximum rpm was 9000 and the engine maximum speed 3000 rpm, the gear ratio would normally be chosen as 3:1.

Looking at the alternator output/speed diagram, we can see that if we run our engine at 1500 rpm, the alternator speed will be 1500 × 3 = 4500 rpm. Its output will be less than 50 amps. If the engine compartment is very hot, then the output may be only 40 amps. If the engine is running at idle rpm, the alternator output is little more than a 'trickle' charge.

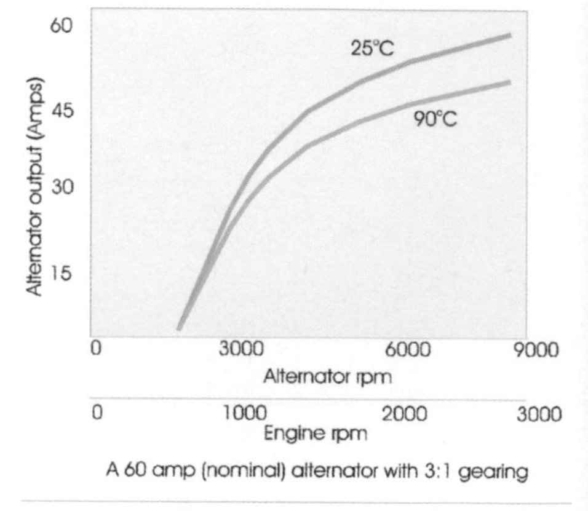

A 60 amp (nominal) alternator with 3:1 gearing

Alternator output

In order to control its output, the alternator needs to know the battery voltage. On many alternators it does this by sensing the output voltage inside the alternator. This is fine if the wiring runs are short, the terminals have good contacts and there's no split-charging diode. This is known as *machine sensing*. Far better is *battery sensing*, where the output voltage is sensed at the battery terminal, because there may be a considerable voltage drop between the alternator and the battery. If this is the case, the battery can never be fully charged.

The alternator must always supply any user of electricity via a battery. The alternator is never connected directly to the 'load'. Provided that the alternator has sufficient output, it will always supply enough amps to satisfy the load. In other words, if the lights need 10 amps, the alternator will supply them with 10 amps.

If the alternator was supplying 10 amps to the lights AND charging the battery, its output would be 10 amps constant for the lights PLUS a diminishing charge going into the battery.

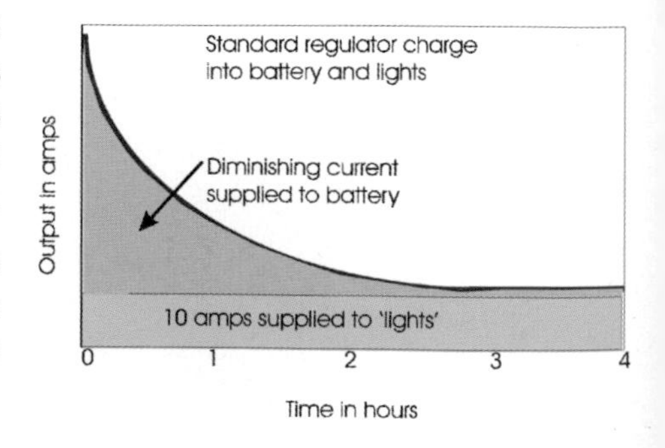

How big should the alternator be?

There are two considerations here: What is the maximum load at any one time? How big is the battery bank?

- Normal electrical loads are always satisfied from the alternator's output.
- This normal requirement may reduce the charge going into the battery.
- The alternator will never satisfy high-current loads such as anchor windlasses and bow thrusters.
- These high loads are supplied by the battery, with the engine running at the same time to reduce battery drain and a fall in battery voltage under this load.
- There's a relationship between the alternator output and the maximum size of battery bank it can realistically charge.

Charging batteries is a complicated science, but as far as we are concerned, it's easy to use a rule of thumb, which says that maximum charging current (amps) needs to be about one-third of the battery capacity (amp hours) for efficient rapid battery charging. In other words, the battery capacity should be no more than three times the alternator output.

A standard 60 amp alternator is therefore suitable for charging a 180 amp hour service battery bank. It's usual to discount the engine start battery, because generally it's pretty full anyway.

If you are going to make heavy use of your battery over a weekend, but then have all week to charge the battery on mains power, the size of your alternator compared to the size of the battery is not so important.

You should never allow your battery to become more than 50% discharged, as discussed in the section on batteries later in this chapter.

Generally, unlike a car, the size of the alternator is dictated by the size of the battery bank. If you increase the size of your battery bank, you may need to fit a bigger alternator or a second alternator.

A bigger alternator

The standard belt drive may need to be upgraded if you fit an alternator of higher output. An overworked belt will shed a lot of black dust and will fail after only a few tens of hours. Its capability will depend on the contact angle of the belt, its section and its tension. A wide, 'poly belt' or multiple belts may be required. I have found that the capability of a single belt can be increased by using a high-temperature belt (as shown in the photograph) in combination with machined pulleys, rather than the standard pressed steel type.

A second alternator

By fitting suitable brackets to the engine, it may be possible to fit a second alternator being driven by a second belt and pulley system. Consult the engine manufacturer to ascertain the maximum side thrust loadings allowed on the crankshaft. Think about how you will connect them electrically, which is dealt with under battery charging later in this chapter.

Starting load

A bigger than normal load on the engine when starting from cold with well-discharged batteries may have undesirable effects such as difficult starting or black smoke after starting. It's almost certain that you will choose to fit a 'smart' regulator if you upgrade your charging system (see Battery charging later in this chapter) and some of these have a time delay before the alternator starts to charge to overcome these problems. This is a desirable feature.

These problems can occur even with a standard system, and I've recently discovered that a modification to the Yanmar 1GM10 fitted with a sail drive leg was made in 1992 but not included in the handbook. To prevent problems with cold starting, the alternator does not produce a charge until the leg's oil temperature reaches 25 °C. The alternator warning light was changed to illuminate when the alternator was producing current, rather than the normal warning function. This resulted in owners having undercharged batteries if the engine was run for only a short period of time, so you need to allow for the non-charging period when charging your battery.

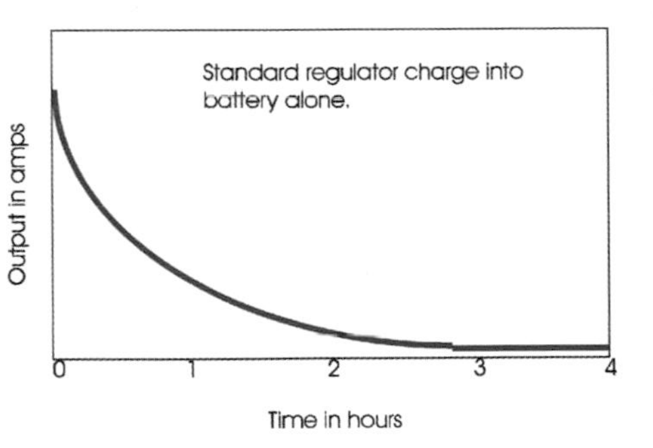

Voltage regulation

A standard 12-volt alternator has a varying output voltage according to the state of charge of the battery. It will charge at a constant voltage of around 14.2 volts with a diminishing charging current. When the charging current drops to a couple of amps, the voltage will reduce to about 13.2 volts. These voltages are doubled for a 24-volt system. As we shall see under Battery charging, this primitive system of regulation will never fully charge our batteries.

A typical alternator

Like an electric motor, an alternator also has brushes, but these are usually hidden inside the casing and a degree of disassembly is required. Sometimes the brush conductor wire will need soldering in place. Alternator brushes suffer more wear than a leisure marine electric motor, but if an engine achieves only 100 hours a year, brush replacement is not likely to be required.

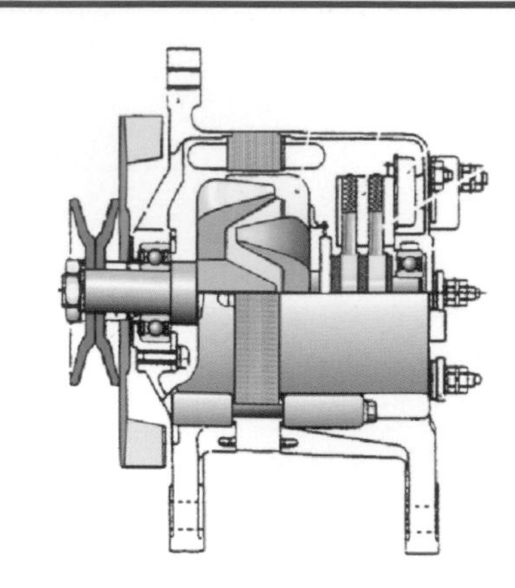

The engine handbook is unlikely to give details of how to change the brushes, but the workshop manual may.

Replacing the brushes

The first job is to switch off the batteries and then disconnect their terminals so that there is no chance of any wire remaining 'live'. If you have the engine wiring diagram, reconnection of the wires will be no problem, but if not, make a sketch showing which wire goes where on the back of the alternator and then disconnect them all.

1. Remove the alternator.

2. Some alternators, like the Lucas, have very easy access to the regulator and the brushes.

3. Remove the two screws holding the cover in place.

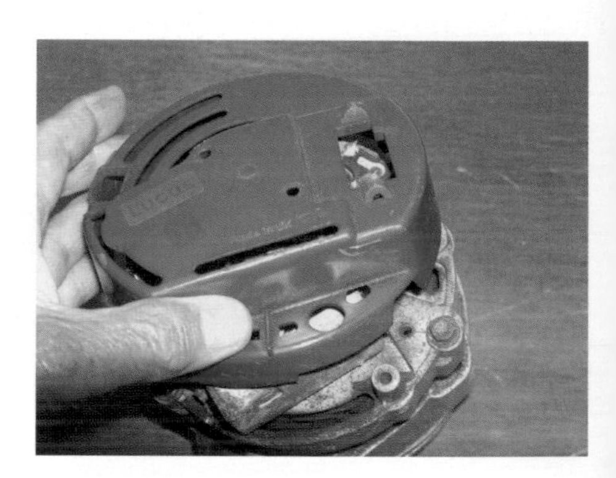

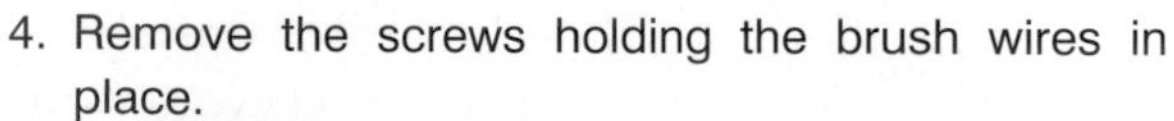

4. Remove the screws holding the brush wires in place.

5. Remove the screws holding the brushes in place.

6. Withdraw the brushes.

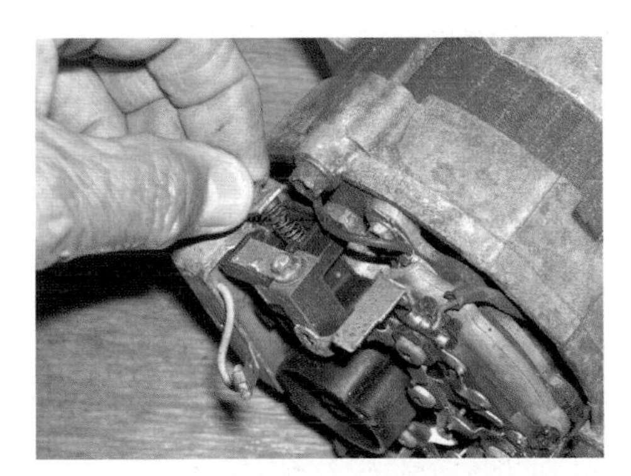

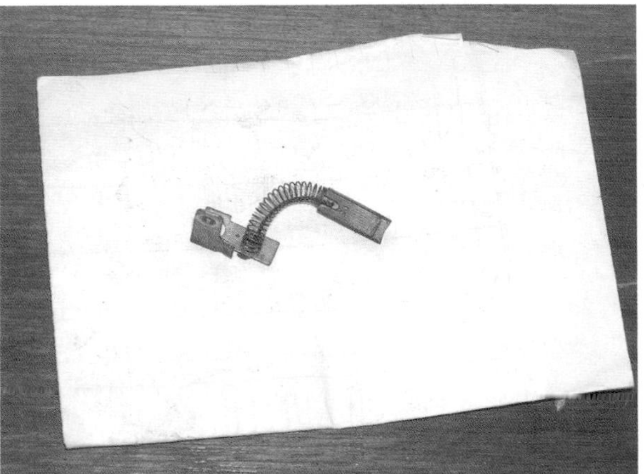

7. Detach the screw holding the regulator wire in place.

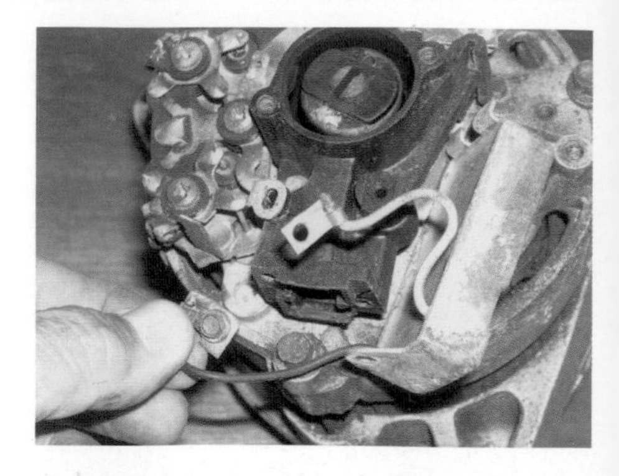

8. Detach the regulator.

Other alternators, like the Hitachi fitted to Yanmar engines, require the alternator to be taken apart to get at the regulator and brushes. The sequence is as shown below when fitting a smart regulator wire to a Hitachi.

Fitting a smart regulator

If a 'smart' regulator is to be fitted, some alternators, such as the Hitachi fitted to Yanmar engines, will need to have the field control wire soldered internally to allow the regulator to operate.

The first job is to switch off the batteries and then disconnect their terminals so that there is no chance of any wire remaining 'live'. If you have the engine wiring diagram, reconnection of the wires will be no problem, but if not, make a sketch showing which wire goes where on the back of the alternator and then disconnect them all.

1. Undo the two bolts attaching the alternator and remove it to the workbench.
2. Remove the five 10 mm nuts at the back of the alternator, noting that three of them have insulating spacers. It might be a good idea to mark the

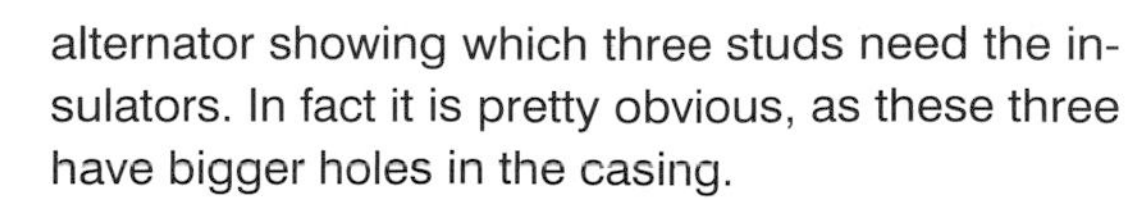

alternator showing which three studs need the insulators. In fact it is pretty obvious, as these three have bigger holes in the casing.

3. The two halves of the casing are clamped together with four long screws and these must now be removed.

4. Carefully prise the front of the alternator casing from the stator (stationary coil assembly), working your way around the perimeter.

5. Separate the alternator unit into its three main parts.

6. The stator rectifier pack and standard regulator assembly also shows the brush assembly, which can now be accessed.

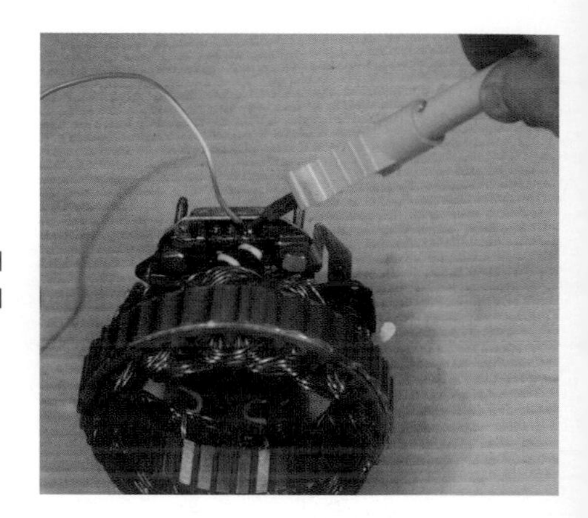

7. Solder the new wire onto the regulator field terminal, as identified in the instructions (it should be marked ‘F’).
8. Slip an insulating sleeve in place.

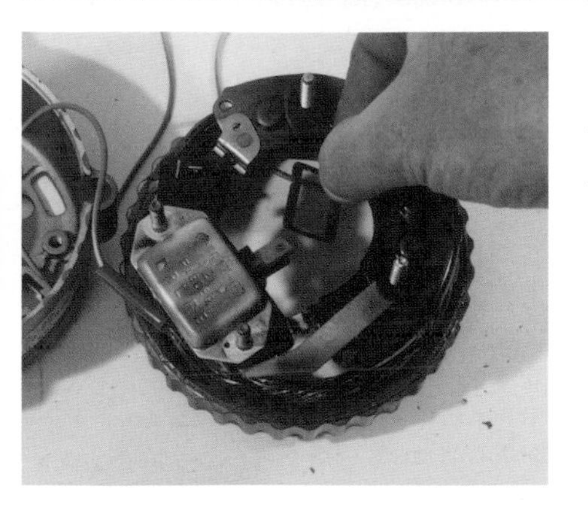

9. Note the carbon brushes and the rectangular brush assembly seal.
10. All is now ready for reassembly.
11. Thread the new wire through the rear of the casing in such a way that it will not foul anything inside and so that you can clip it to a slot with a cable tie. Put the brush assembly seal in place and slip the stator assembly into the rear casing, making sure that the seal is seated correctly.

12. Here, a match has been inserted into a hole in the back of the case to hold the brushes in a retracted position prior to the insertion of the rotor. A match was used for clarity in the picture but is not really strong enough to do the job. A suitable drill bit would be better.
13. Offer up the front casing and guide the bearing and armature gently into place. If any untoward resistance is felt, start again and check that the brushes are fully retracted, as they are easily damaged. Don't force it into place!
14. Withdraw the drill bit and reassemble all the spacers, washers and nuts.

15. Attach a connector to the new wire and you are ready to reinstall the alternator.

16. Tension the drive belt correctly and ensure the attachment bolts are secure. Complete the external wiring using the 'Smart' regulator's wiring diagram and you are ready to sample the pleasures of fully charged batteries.

The first time I did this, the work on the alternator took half an hour. The second time, to take some more pictures, took only 10 minutes!

STARTER MOTOR

The starter motor of a diesel engine needs to be powerful enough to turn a high compression engine at a high enough speed to ensure that the temperature rise due to compression is sufficient to ensure ignition.

The starter motor has a small-drive gearwheel driving the large gearwheel (ring gear) machined onto the diameter of the flywheel. Meshing of the starter motor gear and the flywheel gear takes place only when the starter switch is 'made', so that when the engine is running, the starter motor is not turning.

Starter gear engagement is achieved in one of two different ways:

- a Bendix drive;
- a pre-engaged starter.

Starter/Flywheel ring gear meshing

Bendix drive

Older engines use a Bendix drive. The starter gear has a spiral groove on its inside, and this is mounted on a long shaft with a spiral groove (lead screw) on its outside surface.

When the starter shaft starts to turn, the inertia of the gearwheel inhibits rotation so it 'rushes' along the spiral to engage with the starter ring gear on the flywheel. Its movement along the spiral

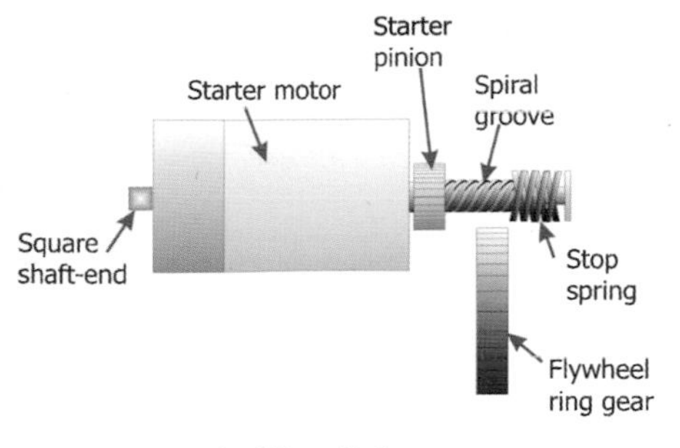

Inertia ('Bendix') starter

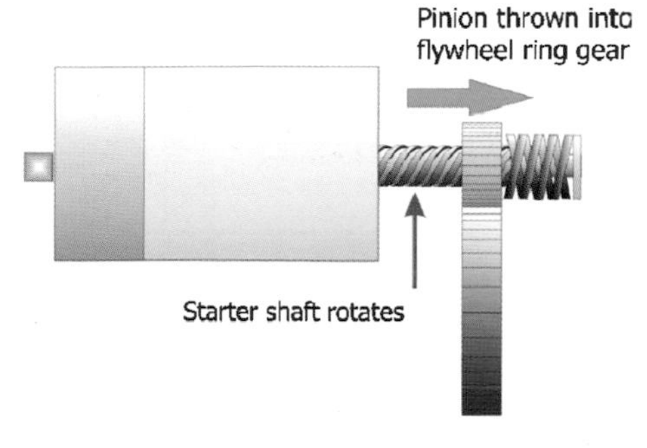

Inertia ('Bendix') starter pinion engaged with flywheel

then halts, so the gear starts to rotate, turning the flywheel.

When the starter switch is released, the starter motor wants to stop, the force between the two gears is removed and the return spring causes the starter gear to disengage and return to its stopped position. It all sounds a bit complicated, but the system works well provided the spiral groove is clean. If it gets dirty, the gear will not move along the spiral, causing the starter gear either to refuse to engage or disengage. However, the constant engaging and disengaging causes wear of both gears on their engagement faces, and this led to the development of the pre-engaged starter.

If the starter doesn't engage, the engine can't be started, the starter motor will just spin rapidly but uselessly.

If the starter doesn't disengage, the starter motor will be forced to rotate rapidly, causing rapid wear. When the engine stops, it can't then be restarted, because with the gears still meshed, the starter won't turn the stationary engine, as it has insufficient power. To overcome this situation, the starter shaft has flats on its end, so that the shaft can be rotated using a spanner to 'wind' back the spiral and gear to retract it. The only proper solution is to clean the spiral groove.

Pre-engaged starter

To prevent wear of the starter pinion and flywheel ring gear as they engage, modern starter motors cause the gears to engage prior to the starter motor turning.

The starter solenoid, when powered from the start switch, has its electromagnetic coil energised. This moves the actuating rod, which, in turn, moves a lever to engage the starter gear with the ring gear. Only then are contacts at the other end of the activating rod 'made', to allow the starter motor to be connected to the battery to rotate the starter

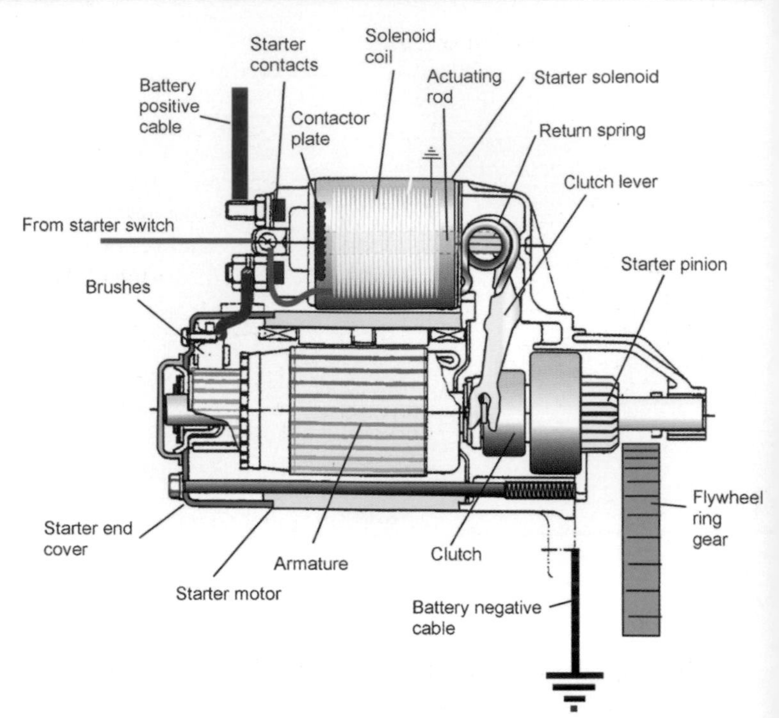

Pre-engaged starter motor

motor. Once the engine fires, the ring gear starts to turn faster than the starter gear and an over-run clutch prevents shock damage to the starter. However, the start switch must be released as soon as the engine starts, to prevent damage due to overspeeding.

Maintenance

Normally the only maintenance required is checking the tightness of electrical connections and the bolts attaching the starter to the engine, which also normally form the electrical negative connection to the starter.

In time, the starter motor may fail, and for the average user, an exchange starter motor is the most

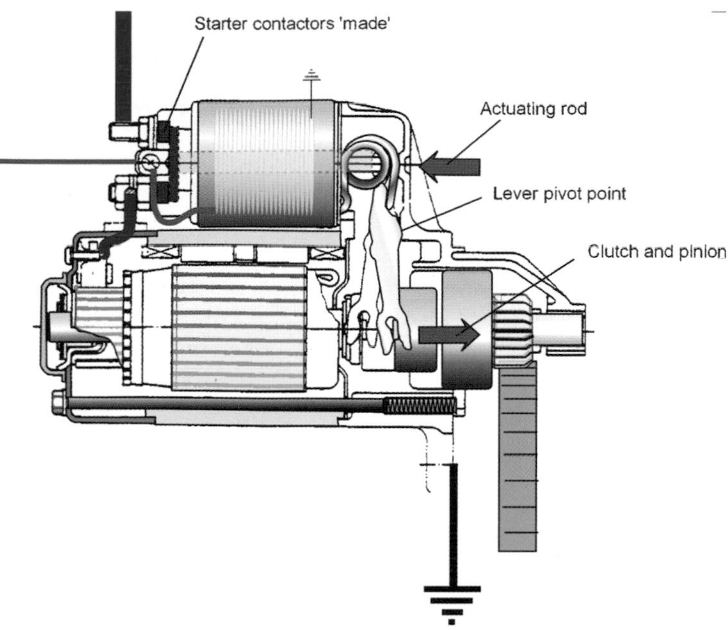

Pre-engaged starter with pinion engaged

likely option, though overhaul is possible. Current is transferred to the rotating coils by carbon brushes, which, in time, will wear, though the starter motor gets so little use that need for their replacement is unlikely.

Checking the engagement mechanism

Operation of the engagement mechanism can be checked if the starter motor is removed from the engine.

1. Disconnect the battery.
2. Remove the starter motor.
3. Hold the motor in a vice and connect the negative lead of a 12 volt battery to the motor's casing.
4. Connect the 12 volt positive lead to the starter 'spade' terminal on the solenoid.

5. Observe the pinion, which should move toward the tapered end of the pinion cover.

Checking operation of the starter motor

If you need to check the rotation of the starter:

1. Disconnect the battery.
2. Remove the starter motor.
3. Hold the motor in a vice and connect the negative lead of a 12 volt battery to the motor's casing.
4. Connect the 12 volt positive lead to the starter's 12 volt positive battery stud terminal on the solenoid.
5. Connect the 12 volt positive lead to the starter 'spade' terminal on the solenoid.
6. Observe the pinion, which should move toward the tapered end of the pinion cover and then start rotating.

Note: as the motor doesn't have to turn the engine, the current will be relatively modest. If the motor casing is not held firmly, there will be a torque reaction that may jerk the motor from the vice.

Replacing the brushes

The brushes may be hidden under a removable band around the motor, behind a plate at the end of the motor or under a deeper rear cover.

The engine workshop manual will cover this operation.

1. First, disconnect the battery.
2. Remove the main cables from the motor and also any signalling wires (solenoid), noting how they should be reconnected. (A starter motor may have no negative cable, as this may be supplied via the engine block to which the motor is bolted.)
3. Remove the motor from the unit.

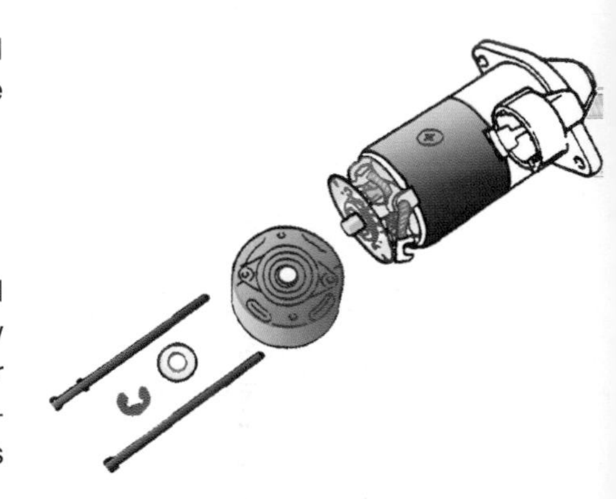

4. Remove the cap at the end of the motor.

5. Remove the ‘C’-shaped retainer on the end of the shaft.
6. Unscrew and remove the very long bolts holding the end cover in place.

7. Remove the end cover to access the brushes.

8. Remove the brushes from their holders – you will probably have to lever the springs out of the way using a small screwdriver.

9. Generally, starter motors have the brush cables soldered in place because they carry very high currents.

10. Unsolder the brush cables with a soldering iron. Use a desoldering tool to remove excess solder from the terminal – you'll need to keep the old solder molten with the iron while you use the vacuum desoldering tool to suck the old solder away.

11. Replace the brushes. This will vary from machine to machine – they will withdraw from the carrier after you have released the pressure on the spring.

12. Reassemble the motor.

DYNASTART

Some old engines are fitted with a *dynastart*. This is a combination of a 12 volt DC generator (a dynamo) and a starter motor.

Modern alternators produce AC current, which is converted internally by diodes into DC current, which can charge the battery and supply the electrical circuits. A dynamo produces DC current directly when it is rotated by the engine, but its power output is relatively low. If a DC current is supplied to a dynamo, it will turn and so can be used as a starter motor, driving the engine through the 'V' belt. Although simple, dynastarts are not very powerful or efficient. Engines such as the early Volvo Penta MD2 initially used dynastarts, but later versions used separate alternators and starter motors, and conversion kits were available.

It's likely that any problems with a dynastart will need attention from a specialist, as exchange units will not be available.

BATTERIES

There are a number of different types of storage battery and people have differing ideas as to which is best. The information here is generally accepted, but the balance may change a little depending on who you talk to. Unless your requirements are extreme, it's often most cost effective to use good quality general purpose batteries, treat them properly and accept that you may need to replace them every four to six years. However, if you dedicate one battery to engine starting only, there's no reason for it not to last more than ten years.

A battery consists of a number of standard 2 volt (lead acid) cells joined in series in one battery case to make up a battery of *nominal* voltage.

Lead acid batteries

The standard 12 volt battery consists of six 2 volt cells. Each cell has a series of positive and negative plates suspended in a solution of sulphuric acid.

The plates are kept apart by separators. Under load, electrons flow from the negative plates to the positive plates, and under charge, the flow is reversed. The amount of electricity that the battery can hold is determined by the surface area of the plates, so a big battery will store more electricity than a small one. But that's not the whole story.

If a battery is required to give a very high current for a short time, such as when starting a diesel engine, the plates must be very thin, so that the stored electricity is available at the surface of the plates very rapidly.

These plates are fragile and if a lot of electricity is taken from them, the plates will buckle. Also, they don't like vibration.

If relatively low currents are required, plates can be much thicker. They are much more robust but won't give high currents, because the electrons can't flow from deep in the plate fast enough.

The physical size of the battery is determined by how much electricity it stores (its capacity) and how quickly the electricity is required. A battery's capacity will decline slowly as it ages. Battery capacity will also vary with temperature, and at 0 °C is only 50% of the nominal capacity compared with that at 30 °C.

Types of lead acid (flooded) battery

- *Engine start battery.* The engine handbook will specify the start battery's capacity in amp hours, and also its current rating (often called cold-cranking amps – CCA). The current rating will be high, typically 400 amps or more, but this current is required for a very short time, and to achieve this, the plates will have to be fairly thin. This battery is not very suitable for supplying the general domestic services of a boat, but will be discharged by only a few percent at each engine start. For this reason, the plates don't have to be very robust, as long as they can withstand the vibration.

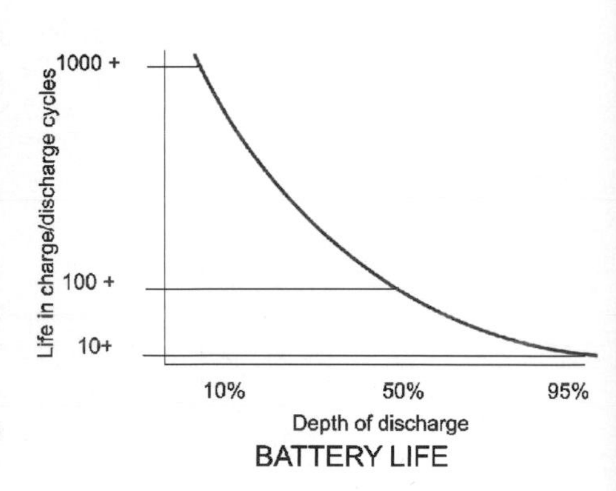

BATTERY LIFE

- *Service battery.* This will need to deliver a relatively low current, mainly between 0 and 15 amps. The plates can, therefore, be thicker and thus more robust. It will need to withstand much deeper discharges between charges if it's going to power the boat's services when the engine isn't running or shore power isn't available. The thicker plates will allow it to achieve this. Even so, the service battery shouldn't be discharged below about 50% of its capacity, because if it is, its life will be severely reduced in terms of how many times (or cycles) you can discharge it.

General purpose batteries

Unless the demands of engine starting are severe, you can use a battery that will fulfil both engine starting and service requirements satisfactorily. It will do neither job perfectly, but the average yachtsman will probably find this type of battery perfectly OK. They are often sold as *marine batteries* or *heavy duty batteries*.

Deep cycle batteries

Deep cycle batteries have much thicker plates and are heavy. The 'deep cycle' adjective doesn't mean that you can discharge the battery much more than 50%, but it does mean that you can do so a greater number of times. Unless the battery is a *traction* battery that is extremely heavy duty and designed for fork-lift trucks, golf carts and the like, it will rapidly be destroyed by fully discharging it.

Some batteries are labelled 'heavy duty' or 'deep cycle' when in reality they do not deserve the title, and it's often difficult to get the facts. One battery company that does publish information is Optima, from Sweden, whose Optima 'Blue Top' marine battery claims to have a BCI/SAE cycle life of 350 cycles of fully charged to 100% discharged. This is an excellent figure.

Batteries should be recharged as soon as possible after being discharged, to prolong their life.

Maintenance-free batteries

Normal batteries lose water from their electrolyte during recharging and need to be topped up regularly with distilled or ionised water. Explosive hydrogen gas is given off in this process.

If the battery has more water to start with, has its charging current restricted and is almost entirely sealed, it will not need to be topped up during its lifetime. These *maintenance-free* batteries have no means of being topped up. They must not be charged as rapidly as a standard battery and thus will take longer to charge, but if you have a standard regulator, you won't notice the difference. A mains battery charger will need to be set for maintenance-free charging. These batteries lose very little charge during storage and should be capable of starting an engine after being stored for 18 months. A standard battery will lose up to 1½% of its charge every week it's idle.

Some so-called 'maintenance-free' batteries have caps to each cell. If it's not sealed, it isn't maintenance-free and will need to be topped up as required.

AGM batteries

Absorbed glass mat batteries have heavy duty plates separated by glass fibre mats that absorb the electrolyte, so that there's no free liquid in the battery. The best are robust, are virtually non-spill and withstand vibration well. They tolerate high charging voltages and have low self-discharge rates and some, such as 'Lifeline', can be used for engine starting as well as being deep cycle 'service' batteries. AGM batteries have good deep cycle/life expectancy.

Gel batteries

The electrolyte is in the form of a gel and the plates are thin to allow the electrolyte to diffuse into the plates. Gel batteries are sealed, so can't be topped up, and the charge voltage must be kept low so that the battery does not gas. Very strict regulation of charging current and voltage is required, and gel batteries take longer to recharge than wet ones.

Nickel-Cadmium (Ni-Cd) batteries

Nickel-Cadmium batteries are very robust and have a very long life, but are very expensive. They can be deep cycled thousands of times, so are very suited to use on a sailing boat. Because they lose the ability to be fully recharged unless they are fully discharged, you really need two service battery banks so that one can be fully discharged before recharging. Each bank is used alternately. Their cost is justified only if you're going to keep the boat for a long time and are going to make heavy demands on your batteries.

Measuring state of charge

The state of charge (how fully the battery is charged) can be determined by means of a hydrometer, which measures the specific gravity of the sulphuric acid solution, otherwise known as the *electrolyte*. The battery needs time to stabilise after charge, or discharge, before the state of charge can be determined.

Generally speaking, using a hydrometer is not convenient on a boat, and the specific gravity of fully maintenance-free batteries cannot be tested. It's much easier to determine the state of charge by measuring the battery's voltage, but to do so the battery must be 'at rest'. In reality, this means that the battery needs to have been neither 'on charge' nor discharge for around three hours and, again, this is not very practical.

Meters that claim to be battery 'state of charge' meters just cannot work unless the battery has been at rest for three hours or so. In reality, they are just voltmeters with a different scale.

To get some idea of the instantaneous state of charge of your battery, you can use a voltmeter and an ammeter in conjunction. A fully charged battery at rest has a voltage of about 12.8 volts. Fully discharged, its 'at rest' voltage is about 11.8 volts. Both of these figures are reduced if a load is applied. In fact, try and start the engine with a fully charged battery and the voltage will drop to around 10 volts. Do that with a flat battery and the voltmeter will drop close to zero.

12.4V

Battery 100% state of charge

If you observe the ammeter and the voltmeter together, the table shown will give a fairly good indication of the battery's state of charge, even while it's being used.

BATTERY STATE OF CHARGE	BATTERY VOLTS			
	RESTED	0 AMPS	5 AMPS	10 AMPS
100%	12.8	12.5	12.4	12.2
90%	12.7	12.4	12.3	12.1
80%	12.6	12.3	12.2	12.0
70%	12.5	12.2	12.1	11.9
60%	12.4	12.1	12.0	11.8
50%	12.3	12.0	11.9	11.7
40%	12.2	11.9	11.8	11.6
30%	12.1	11.8	11.7	11.5
20%	12.0	11.7	11.6	
10%	11.9	11.6		
FLAT	11.8	11.5		

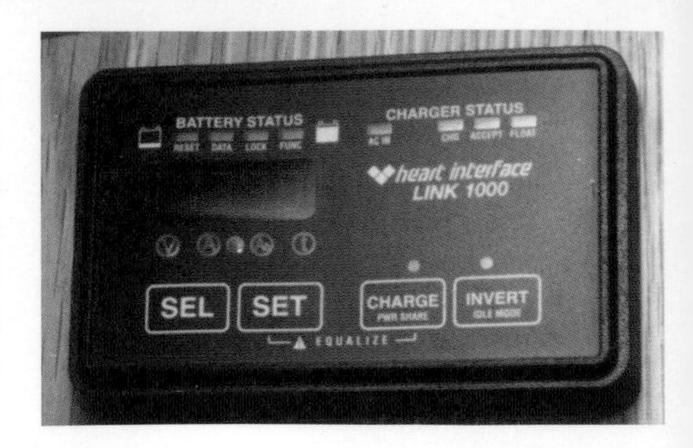

Using an amp hour meter

Another way is to use a sophisticated electronic circuit that integrates the current that has been flowing for a given time, does the sums and tells you how much has been used. With even more refinement, these meters can tell you how much charge you have put back into the battery by allowing for the efficiency of the charging process. They are not absolutely accurate, but are very effective (although expensive).

Sulphation

Sulphation is a natural process during discharge and recharge cycles, where a layer of lead sulphate is built up on the battery's plates and this layer reduces battery performance. Sulphation can be removed only by bringing the battery back to a full state of charge, and becomes a serious problem in deep cycle batteries that rarely get fully charged. Initially soft and porous, this layer hardens with time and, once hardened, it can't be removed, rendering the battery useless. For this reason, the battery should always be left fully charged.

Some charge regulators/battery chargers have an *equalisation* or *conditioning* setting that can be used monthly for those batteries that are deep cycled regularly. Because high voltages (up to 16 volts) are used, all electronics must be disconnected during equalisation and must not be used for gel batteries.

Self-discharge

Batteries not in use will discharge themselves over a period of time. Traction batteries self-discharge at as much as 1% per day – the higher the temperature, the higher the rate. General purpose batteries are better, and sealed lead acid batteries lose only 0.1% per day.

Because of this self-discharge, sulphation will occur and monthly recharging of non maintenance-free batteries is required when the batteries are not in use.

Ageing of batteries

Lead acid batteries will last an extremely long time if they are never discharged more than about 5%. My last engine start battery was still going strong after

12 years. Service batteries, because of their regular cycling, will slowly suffer from irreversible sulphation and their effective capacity will fall. Regularly fully charged and never discharged below 50%, you may expect five or six years, maybe more. Mistreated batteries may not last two seasons.

If one cell fails, this will pull the voltage down, not only on that battery, but also on the whole bank. If the battery bank voltage has fallen to 12.5 volts or so, after being fully charged and rested for 12 hours, you can suspect a failure of one cell. There are three ways of checking which battery has the bad cell.

- Measure the specific gravity of each cell with a hydrometer. The bad cell will have a much lower reading.

- Disconnect all the batteries. Wait about 12 hours and measure the voltage of each battery. The bad battery will still be 12.5 volts or less, the others should have recovered a little.

- Disconnect all but one battery at a time and use it to turn the engine over. You'll need to prevent the engine starting by setting the stop control to stop. The battery that drops below 9.5 volts has the bad cell.

If the whole battery is suffering from sulphation, its real capacity will be reduced. To test this, discharge the fully charged battery by using a number of lights of known wattage for long enough to discharge it by 25% of its nominal capacity. Now measure its specific gravity or its 'at rest' voltage to determine its actual state of charge. The difference between the actual state of charge and the nominal 75% indicates the reduction in capacity.

So, for a nominal 200 amp hour battery, you want to remove 50 amp hours to reduce it to 75%. Four 10 watt bulbs on a 12 volt system would take 3.33 amps and would need to be run for 15 hours to remove the 50 amp hours. If, after 12 hours at rest, voltage is now 12.3 volts, the actual state of charge is only 50%, rather than the nominal 75%. The lights have removed

50% of the capacity instead of the hoped for 25%, so the battery now has only half of its nominal capacity and it's time it was retired.

Topping up lead acid batteries

Lead acid batteries (excluding genuine maintenance-free) need to be topped up with distilled water. Batteries that are worked hard need attention more frequently, so start off by checking monthly and adjust the time period as necessary.

1. Wipe the battery tops with a clean cloth.
2. Remove the stoppers of each cell – these may be screw-in types or lift-off types.
3. The liquid electrolyte should cover the plates by about half an inch (10 mm).
4. Top up, if necessary, with distilled water – de-ionised water from a car accessory shop is the cheapest.
5. Some batteries have a tube that reaches down to the correct top-up level descending from the filler neck – top up to the bottom of the tube.
6. After you've finished, wipe the batteries down with a solution of bicarbonate of soda to neutralise any acid.
7. Note the date that you topped up.

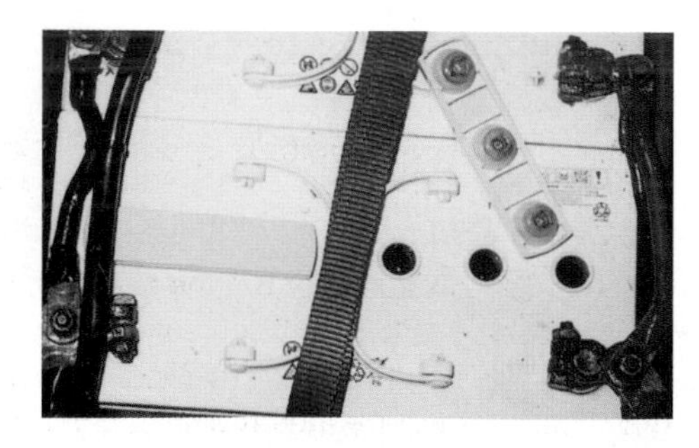

Battery charging

On a car, the alternator is used to supply the load, such as the lights and heater. The only battery charging it's designed to do is replenishment of engine starting load, which is pretty small, so if the battery gets really discharged, the car alternator is not really able to cope. This would be true on a boat that makes demands on its battery only when the engine is running. On most sailing boats, the battery will need to be recharged and this is true also on some motor boats.

Unfortunately, the charge regulator on a marine engine is just the same as on a car. The alternator output is regulated in a very rudimentary manner, and its output current is forced to drop sharply after a very short time in order not to overcharge the battery. So your battery never gets fully charged, and once your battery gets moderately discharged, you can't get it back above 70% in any reasonable engine running time.

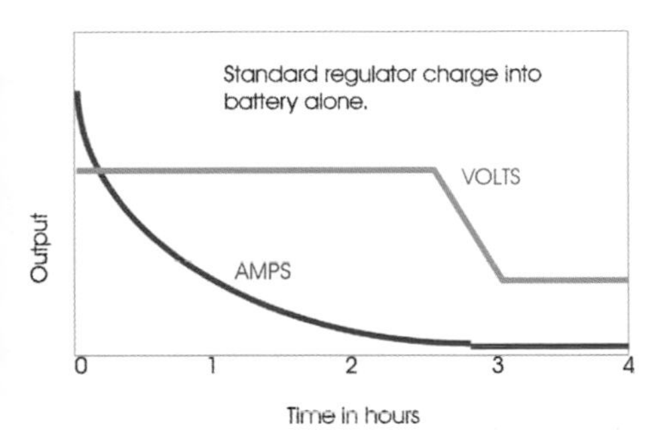

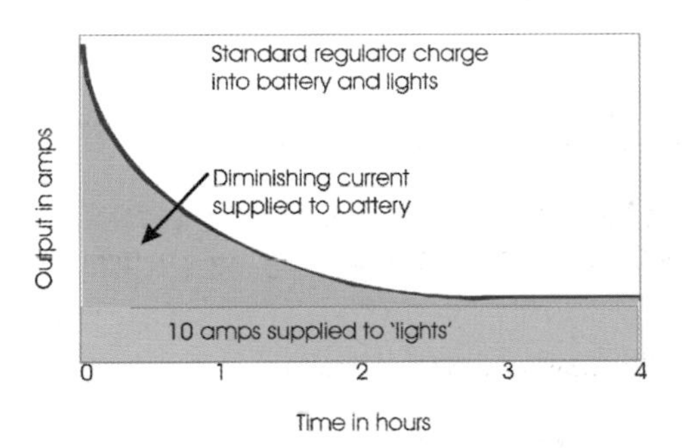

The charge entering the battery is about 90% of the area under the charging curve, because of the inefficiency of converting the in-going current into battery charge. In 1½ hours of engine running, you will be lucky to put 30 amp hours back into your battery with a 60 amp alternator. After 1½ hours, the in-going current will be little more than a 'trickle charge'.

If most nights will be spent connected to a shore power supply, this rudimentary charging system will be entirely adequate.

'Smart charge' regulator

The charging power of your existing alternator can be increased dramatically by fitting an external *smart* regulator. Overcharging is prevented while, at the same time, a high charging current is maintained for as long as it's safe, before the current tails off.

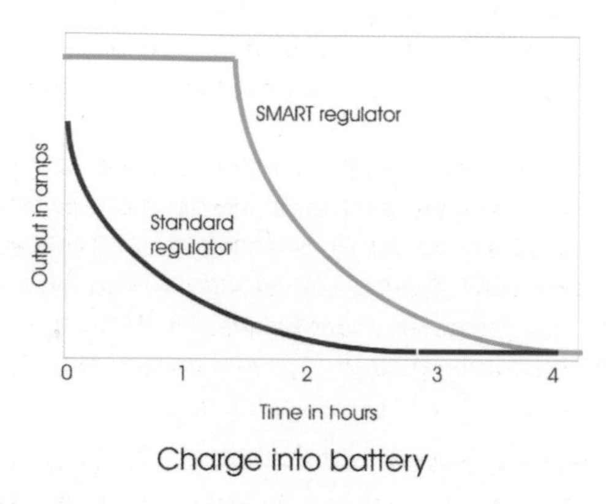

Charge into battery

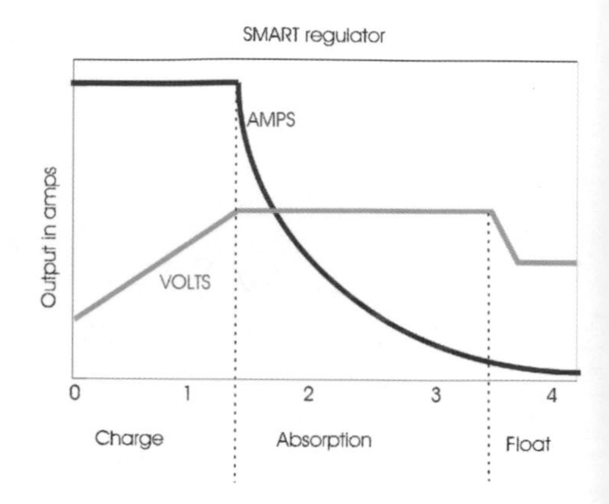

In any given time, the area under the charging curve is much greater, and the trickle charge stage is reached after a much longer time. This system will allow a battery to reach a 90% state of charge in a realistic time. In 1½ hours of engine running, our 60 amp alternator will put around 75 amp hours back into the battery, a considerable gain.

Whatever the type of regulator, the first hour of charging will give the greatest gain, because, after that, output is falling and it's not very effective to continue charging beyond 1 to 1½ hours unless the engine would be running anyway.

However, when the battery is not fully charged, sulphation will occur, so the battery needs to be brought up to a fully charged state as frequently as possible. You need to balance maximum charging gain with sulphation and battery life to get the best possible cost effectiveness.

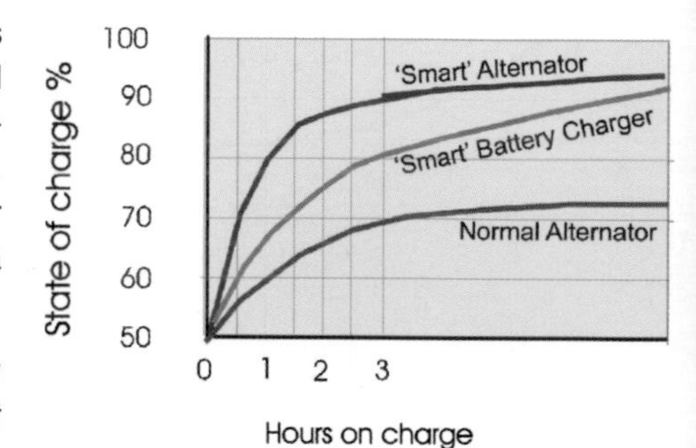

TIME TO CHARGE A BATTERY

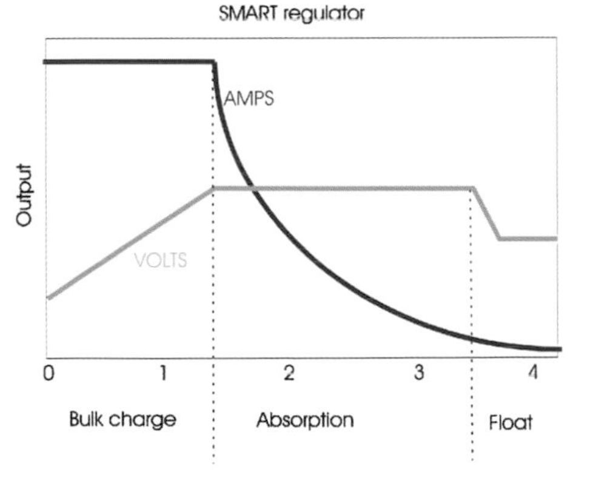

Charging from a mains battery charger

Cheap mains battery chargers recharge a battery in the same manner as a standard marine regulator. Unless you have plenty of time (days rather than hours), these are of little use on a boat.

Proper marine mains battery chargers work in the same manner as a smart regulator. They have multiple stages of charge programme to enable the battery to be charged efficiently, i.e. to bring the battery to as high a state of charge as possible in the minimum time.

- *Bulk charge* maintains a constant current as the battery voltage increases, up to the point at which 'gasing' occurs – typically 14.4 volts. Above this voltage the electrolyte begins to break down into hydrogen and oxygen gases, causing loss of electrolyte. This varies according to the type of battery and is normally set by the user via a switch.

- *Absorption charge* maintains the voltage close to the gasing point and the charge current drops off as the state of charge rises until the battery is fully charged.

- *Float charge* keeps the battery topped up and compensates for the battery's self-discharge. This float voltage is typically 13.5 volts.

Good battery chargers may be connected indefinitely without risk of overcharging, but do read the instructions.

At any time, the charger is able to supply any DC circuit that may be switched on, but this will prolong the time required to recharge the battery fully, especially during the bulk charge phase.

Some manufacturers refer to additional phases. In reality, they are adding the inter-phase switching periods and the time supplying external loads as additional phases.

Most good chargers have the ability to run a desulphation programme, known as *equalisation*. This can harm the battery if used incorrectly, so make full reference to the charger's instructions, especially with regard to battery type, frequency of use and time the high voltage is applied.

If the requirement is to 'top up' your battery overnight, then its output in amps is as important as its programmed charge ability. For this you need to know how much you are likely to have discharged your battery or the capacity of the battery bank. Generally, these two will be closely linked, so it's normal to match the charger output to the capacity of the battery bank.

A good starting point is that the charger's output should be about 10% of the battery capacity (ignore the engine start battery). If high battery loads will be applied at the same time as charging, you may need to add this extra load to the output of the charger.

If you have a 400 amp hour battery bank, then a 40 amp charger is appropriate. If there is a continuous requirement for a further 10 amps, for a fridge and a cabin heater for example, this 10 amps should be added to give a 50 amp charger.

The Gearbox

The gearbox, often referred to as the *marine gear*, has three purposes:

- to reduce the maximum speed at which the propeller rotates;
- to enable both forward and astern rotation of the propeller;
- to enable the propeller to remain stationary while the engine is running.

Unlike a car gearbox, there's only one forward gear, although one manufacturer produces a two-speed gear for high-speed motor cruisers, and the clutch is contained within the gearbox.

SPEED REDUCTION

Modern engines run at too high a speed for normal operation of the propeller. The reduction gear ratio is normally between 2.0:1 and 3.0:1 and many manufacturers have a number of different ratios, so that the engine can be tailored to the boat's requirements. If you are planning to put a new engine in your boat, it's best to discuss the required gear ratio with the engine supplier.

FORWARD AND ASTERN

The input shaft to the gearbox is separate from the output shaft. It can drive the output shaft by either of two sets of gears to give forward and astern, and the output shaft is connected to the appropriate set of gears using one of two separate clutches. It is these clutches that are operated by the 'gear selector' rather than the gears themselves, as is the case in a car gearbox.

NEUTRAL

If neither of the two clutches is engaged, the propeller shaft doesn't rotate, to give 'neutral'.

GEARWHEELS

The meshing gear teeth must be identical in shape and size, even on gearwheels of differing diameter. The gear ratio of a pair of gears is determined by the number of teeth on each. So, if the input gear has 10 teeth and the output gear has 20 teeth, the gear ratio is twenty divided by ten, i.e. 1:2. The output rotates at half the speed of the input.

Most gearboxes state the ratio with the input speed first and the output second, so that a ratio of 3:1 indicates that the input (engine) runs three times faster than the propeller shaft. Often, the reverse gear ratio is different from the forward ratio and is caused by the design of the gearbox, requiring that the reverse idler gearwheel has to be smaller than the normal forward gearwheel.

TYPES OF GEARBOX

There are several types of gearbox in general use:

- two-shaft;
- layshaft;
- epicyclic;
- bevel gear.

All may be operated manually or hydraulically.

Two-shaft gearbox

The simplest and most compact gearbox has two shafts – the input shaft and the output shaft. The input

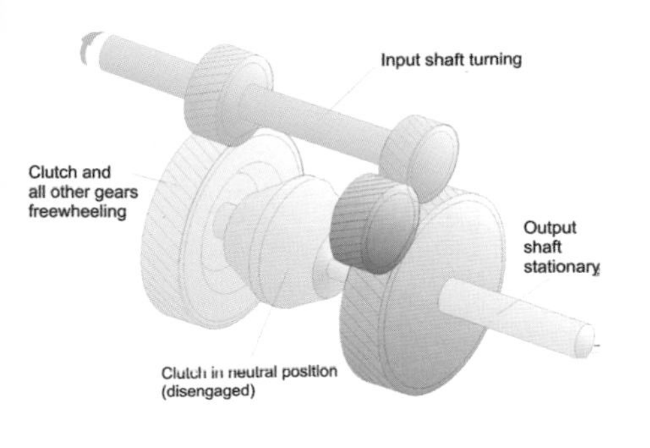

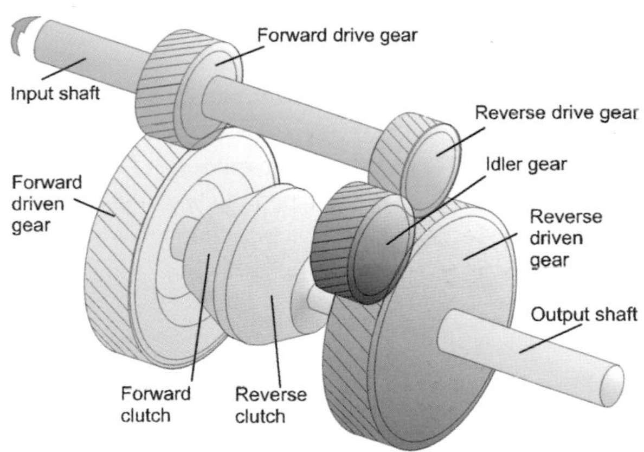

Two-shaft gearbox

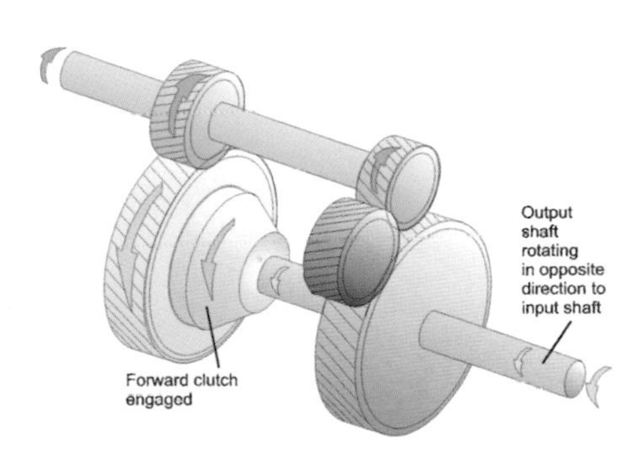

Two-shaft gearbox – forward gear

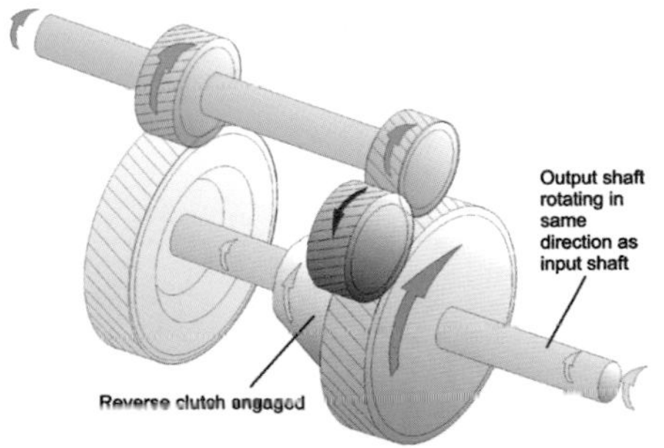

Two-shaft gearbox – reverse gear

shaft drives the forward gear directly and the reverse gear via an idler gear. A cone clutch or plate clutches are moved along the freely rotating output shaft from the centre, neutral position, to link either the forward or reverse gears to the output shaft. The clutch mechanism can be designed so that as the clutch engages, more force is provided automatically to ensure full engagement. This simplifies the design but requires accurate adjustment of the mechanism, which normally adjusts automatically for any wear in use. The downside of this *servo mechanism* is that if you select 'astern' when sailing to stop shaft rotation, it's often difficult to select the lever to neutral. In this case it's necessary to start in gear and as the engine fires, ease the lever into neutral.

Larger versions of the two-shaft box can have the clutch mechanism operated hydraulically. The oil is pressurised by an oil pump, and two control valves, controlled by the gear selector, direct the fluid to the appropriate clutch. The oil serves both as lubricant and hydraulic fluid and is often cooled by an oil cooler in the engine-cooling circuit.

In a two-shaft box, the output shaft exits the box lower down than the axis of the input shaft and rotates in the opposite direction to the engine in forward gear.

Layshaft gearbox

The principle of operation of a layshaft box is similar to the two-shaft box, but a separate, intermediate shaft – *the layshaft* – is used. One clutch is mounted on the input shaft and the other on the layshaft. The output shaft rotates in the same direction as the engine in forward gear. These boxes normally use hydraulic actuation of the clutches.

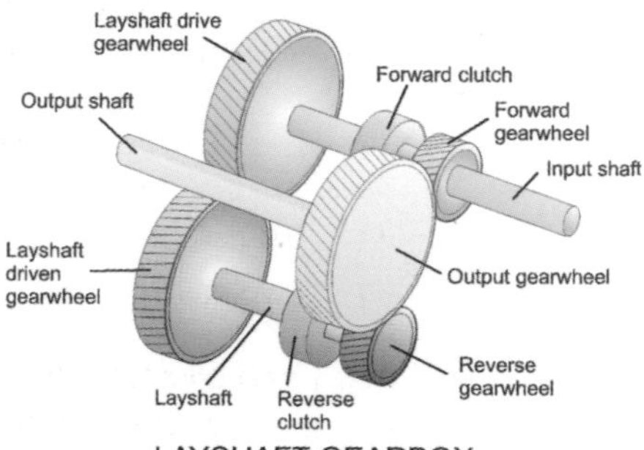

LAYSHAFT GEARBOX

A typical layshaft gearbox is shown in the three photographs – the gearbox has been 'exploded' to make its construction clear.

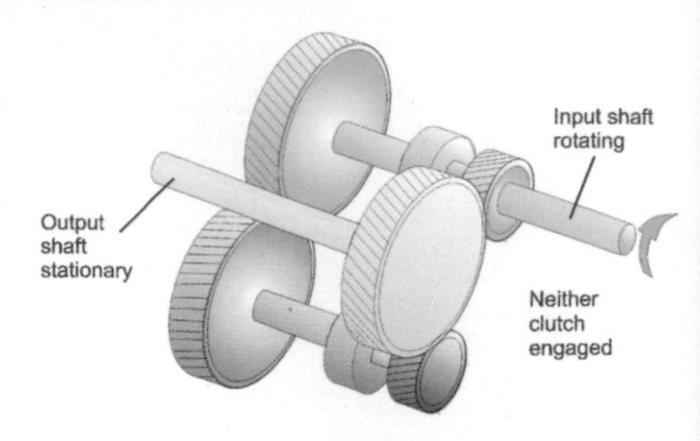

NEUTRAL

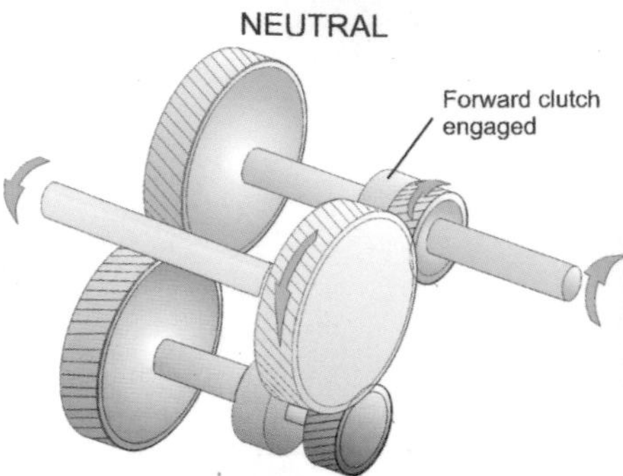

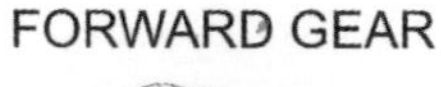

FORWARD GEAR

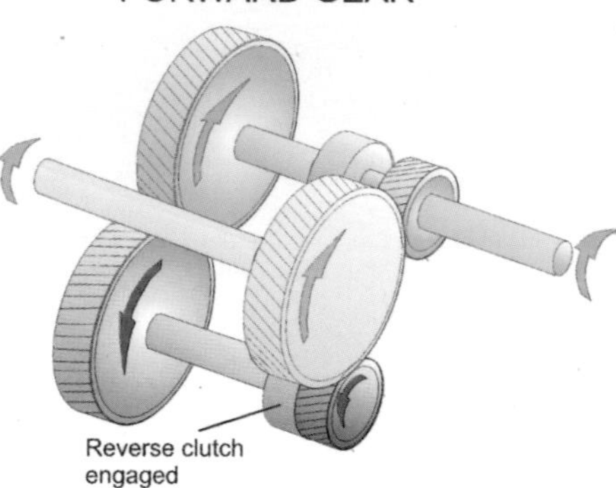

REVERSE GEAR

Epicyclic gearbox

Not only is the principle of operation of this type of gearbox difficult to explain, but so is producing a simple diagram of its operation!

At the end of the input shaft is a gearwheel, known as the *sun wheel*. Around this gearwheel are three equally spaced pairs of *planetary* gears, one of each of the three pairs being meshed with the central input *sun* gear. The three pairs of planetary gears are mounted on a single carrier. The other gearwheel of each of the three pairs is meshed with the gear machined on the inside surface of an annular gearwheel.

Input shaft
Annulus gear
Carrier
Sun gear
Output shaft
Planet gears

The clutches are omitted for clarity

THE PARTS OF AN EPICYCLIC GEARBOX

Situation 1

If the input gear and the carrier on which the planetary gears are mounted are all locked together, the input shaft and the carrier rotate as one, and if the carrier is locked to the output shaft, the propeller shaft rotates in the same direction and the same speed as the input shaft.

Situation 2

If the carrier is locked to the gearbox casing, preventing any rotation, the planetary gears rotate and transmit that rotation to the annular gear, which rotates in the same direction as the input shaft but at a lower speed. The speed of rotation will depend on the relative sizes of the annulus and the input gear. The annulus gear, if now locked to the output shaft, rotates the propeller at a lower speed than the engine rpm and in the same direction.

Situation 3

If the carrier is free to rotate and the annular gear is locked to the engine casing, the carrier rotates in the opposite direction to the input shaft and at a lower speed, according to the relative sizes of the input gear and the planetary gears.

Forward gear

Situation 2 normally provides forward gear. The carrier is locked to the casing by a clutch and the annular gear to the output shaft by another clutch, both controlled by the gear selector.

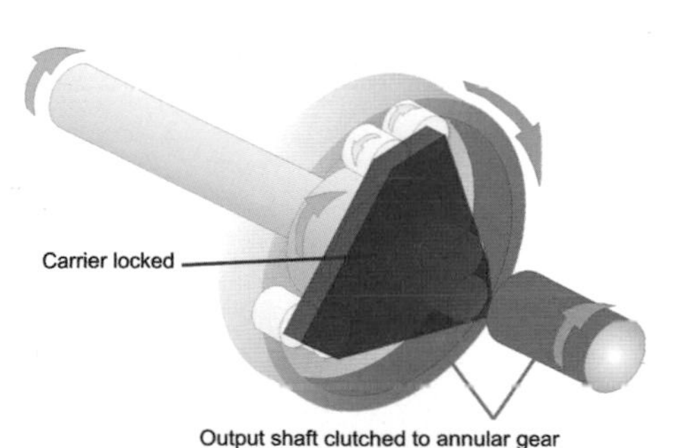

FORWARD GEAR

Reverse gear

Situation 3 will provide reverse gear by using a clutch to lock the annular gear to the casing, with the carrier now being locked to the output shaft by another clutch. Both clutches are controlled by the gear selector as before.

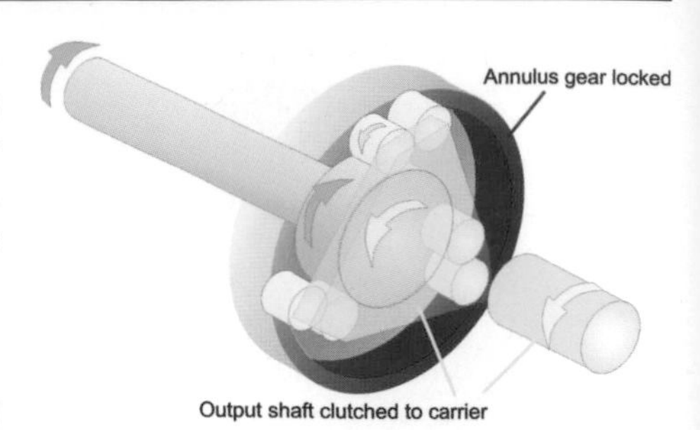

REVERSE GEAR

Neutral

With all the clutches released, the output shaft will not be driven. However, under sail with the engine stopped, there is no way of operating the clutches to prevent shaft rotation, and the freewheeling prop could then cause the gearbox oil to overheat. A shaft lock should be fitted to a sailing boat with an epicyclic gearbox.

Clutch operating forces can be high, and it's normal for epicyclic gearboxes to have hydraulic actuation and for an oil cooler to be fitted.

NEUTRAL

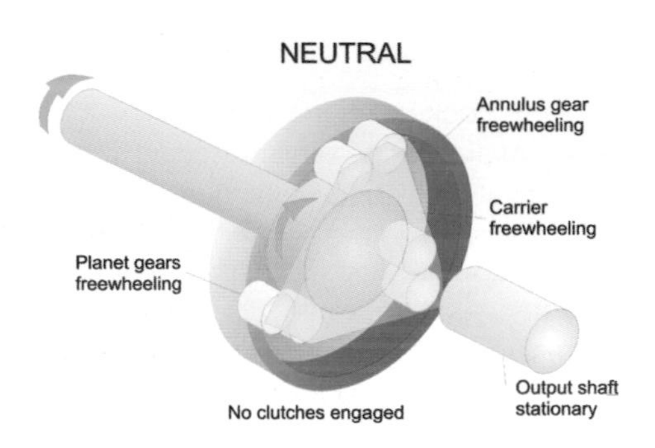

Other types

Sometimes situation 1 is used to provide forward gear, but in this case a reduction gear is used after the epicyclic part of the box to drive the output shaft, otherwise the gear ratio would be 1:1.

Bevel gears

Some gearboxes use *bevel gears*. The use of bevel gears allows the input and output shafts to be at an angle to one another. Sail drive, outdrive and Z drive units all embody the use of bevel gears, as changes of rotation axis direction are required.

BEVEL GEARS

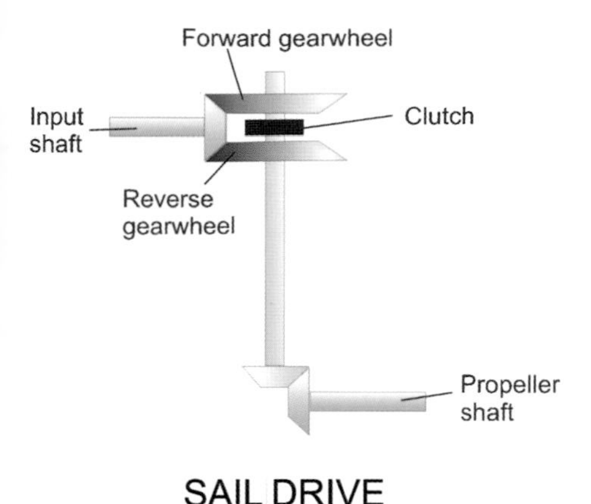

SAIL DRIVE

Sail drives and Z drives have similar principles of operation, with a vertical drive shaft in the leg to connect the input shaft of the gearbox and the propeller shaft at the bottom of the leg. The clutch engages either the upper or lower bevel gear to provide forward or reverse gear.

Volvo Penta used bevel gears for many years in their standard gearbox on their compact engine range. This incorporated a seven degree downangle as standard, but the competition of other engines using much cheaper two-shaft gearboxes spelled the end of the MS2 gearbox.

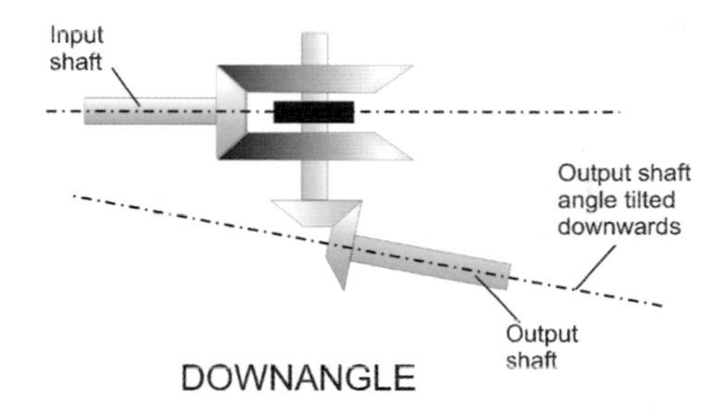

DOWNANGLE

LUBRICATION

Gearboxes may use the same grade of oil as the engine, a different grade of oil or automatic transmission oil (ATF). Some early marine gears actually shared the engine's lubrication oil. Use the specified oil and change it at least as often as the handbook requires.

MAINTENANCE

Other than checking the oil level and changing the oil as indicated in the handbook, there's no other routine maintenance required.

Check the oil level weekly to ensure there's no loss of oil due to leakage or change in colour.

- Low oil level is an indication of leakage, as gearbox oil is not 'burned'. External leakage would be indicated by oil accumulating in the bilge. Internal leakage into the heat exchanger, if fitted, would not be obvious, as most systems are cooled by raw water, so leaking oil would go out through the exhaust.

- A darkening of the oil indicates overheating. This may be due to excessive friction in the gearbox, slipping clutches or blockage of the heat exchanger if fitted.
- A white emulsion indicates water contamination. This is likely to be leakage from the gearbox cooling jacket or heat exchanger if either is fitted, or from the underwater seal in sail drive and Z drive units.

TROUBLESHOOTING

Modern gearboxes are sturdy and generally give little trouble, however, they rely on their clutches being adjusted properly. A slipping clutch will cause overheating and, eventually, loss of drive.

Most clutches are self-adjusting.

- Mechanically selected clutches require that the operating cable moves the selector lever to the fully engaged position, and stretching of the operating cable can cause incomplete selection of the clutches, resulting in clutch slip. Many handbooks will not give details of this adjustment, so you may need the workshop manual or a dealer's help.
- Hydraulically operated clutches use the selector cable to operate the hydraulic selector valve, and, generally, adjustment is not so critical.

Loss of drive

If you experience loss of drive, you'll need to check the rotation of the prop shaft and the movement of the gear selector arm on the gearbox.

1. Check that the selector lever moves to forward and reverse when the 'Throttle/Gear' lever is moved to forward and reverse.
2. If it does, check that full travel occurs.
3. If the selector lever doesn't move, the problem is the cable or cockpit gear selector mechanism.
4. On a 'dual station' system, is the other selector in neutral or the station selector control set appropriately?

Warning!

Ensure that nothing you are wearing or your hair can get entangled in the rotating machinery. Failure to heed this warning can be fatal!

5. If full travel on the selector lever occurs, check that the output coupling rotates.
6. Does the shaft turn as well? If not, you'll need to check the coupling bolts.
7. If the shaft rotates, the problem is the propeller. Is it fouled, loose or missing?

Volvo Penta MS2 gearboxes fitted to Volvo Penta 2000 series engines

There are a large number of these engines in service. Engines prior to the engine number 2300059127 have been found to suffer a tendency for the splined drive shaft to wear, causing failure to transmit any drive. Volvo Penta modified the drive to the gearbox by fitting a cushioned carrier drive to the back of the flywheel on engines from number 2300059128 onward, as a preventive measure.

Examination of the spline requires the engine to be moved forward 100 mm after the gearbox attachment bolts have been removed. The damaged spline seems always to be accompanied by a dry, 'reddish' appearance, and the splines will be seen to have their 'flats' worn away.

DB Marine, Cookham Bridge, Cookham on Thames, Berks, SL6 9SN, Tel:(01628) 520564 have an excellent repair scheme for this problem, which will prevent recurrence.

Drive Belts

A rubber drive belt is used to drive the alternator (or, on older engines, the dynamo). Correct tension is important – too tight and bearings will suffer, too loose and slip will occur. Slip will cause overheating of the belt, leading to its premature failure, and also, where the tachometer is powered from the 'W' connection on the alternator, rpm will under-read. A broken drive belt will stop battery charging, but the engine will run.

On some engines, the water pump is also belt-driven. Failure of the drive belt will cause rapid overheating of the engine and prevent continued running.

Belt pulley alignment is also important for the longevity of the belt, although this will normally be a problem only if the pulleys or alternator have been changed.

ADJUSTING THE TENSION

The method of belt tension adjustment on the majority of engines is pretty crude.

1. Slacken the nut and bolt on the adjustment arm.
2. Slacken the main attachment pivot bolt.

3. Place a long, stout lever between the engine and the alternator so that you can force the alternator outwards to tighten the belt.
4. While still maintaining pressure, tighten the adjustment and pivot bolts.
5. Check the tension and readjust as necessary.
6. Adjusting pump belt tension is carried out in a similar manner.

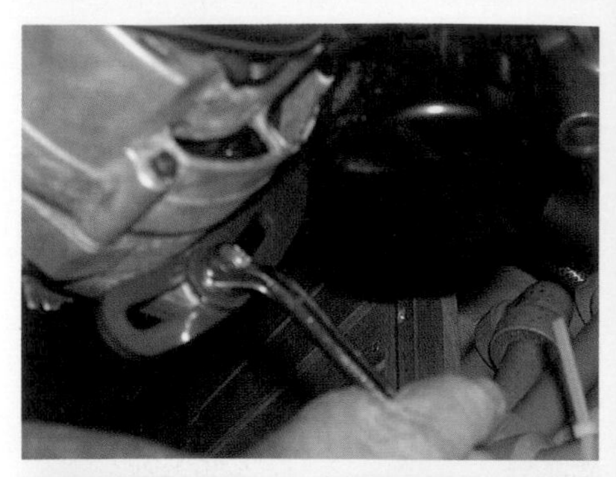

MEASURING THE TENSION

Without a special gauge, measuring tension can be achieved only approximately. However, a rule of thumb method is sufficient for normal use on belt lengths up to about 450 mm.

There are two methods you can use – the *push method*, for short, twin or wide belts, and the *twist method*, for normal belts.

The push method

1. Push the belt at its mid point.
2. Deflection should be about 10 mm.

The twist method

1. Grip the belt at its mid point, between the thumb and forefinger.
2. Twist the belt.
3. Tension is correct if it twists approximately 90 degrees.

SHORT BELT LIFE

If you experience very short belt life, especially if you've fitted an alternator smart regulator or a high-power alternator, try fitting a heavy duty or high-temperature toothed drive belt.

Throttle and Gear Selection

Strictly speaking, the gear selector should be called the clutch and the throttle should be called the speed control. However, everyone knows what we mean if we use the usual terms.

When you move the lever from neutral to ahead or astern, you should move it positively to engage the clutches quickly, this prevents clutch slip with its attendant wear. Don't 'ease it' into gear.

When moving the lever from ahead to astern (or vice versa), pause in neutral rather than push the lever right through. I count 'one and' as I pause in neutral. This is much kinder to the gearbox. Electronic controls do this automatically – you can't rush them.

TWIN-LEVER SYSTEMS

Mechanically, the simplest form of throttle control has a single lever operating a 'push/pull' cable attached to the speed lever of the engine. Similarly, a separate gear selector lever operates the gearbox gear selector lever by a 'push/pull' cable. These are common in the USA, but less so in Europe. Their mechanical simplicity is a distinct advantage, and selection from neutral to ahead and neutral to astern is much less 'notchy'. For those not familiar with them, they take a bit of getting used to, especially when you have twin engines, as you end up with four levers, but they're much easier to 'nudge' along at a very slow speed.

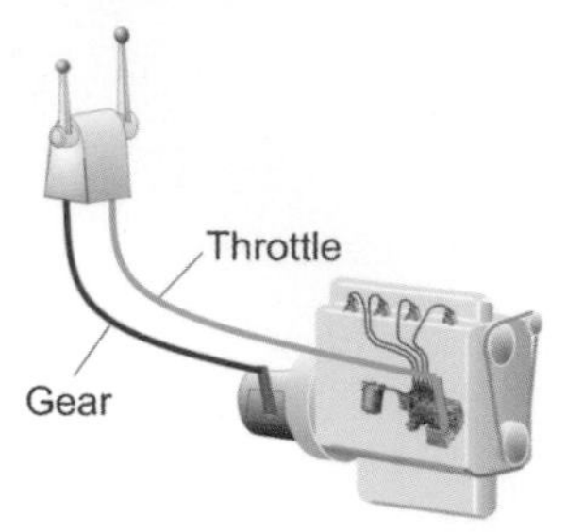

Throttle and gear controls to engine

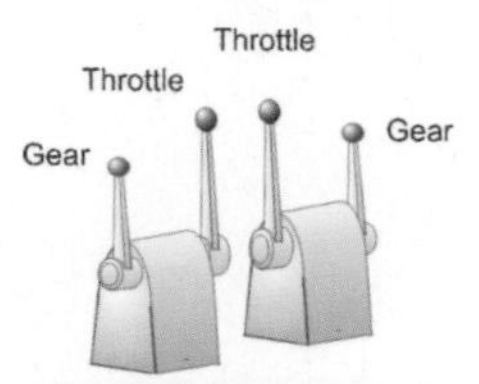

Separate controls
– twin engines

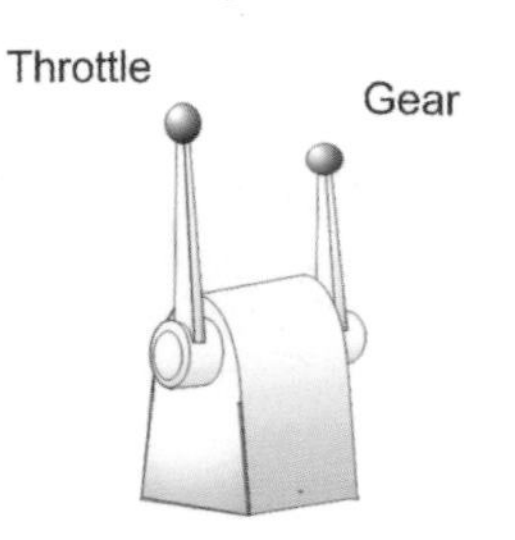

Separate controls
– single engine

SINGLE-LEVER CONTROLS

Many boats have a *single-lever* control. A single lever is pushed forward from 'neutral' to engage forward gear, and further movement forward increases engine rpm. Similarly, rearward movement of the lever engages astern gear, and then further movement rearwards increases engine rpm. The 'throttle' is attached to a control unit incorporating the two functions to operate two separate 'push/pull' cables connected to the engine and gearbox. It's essential that the mechanism

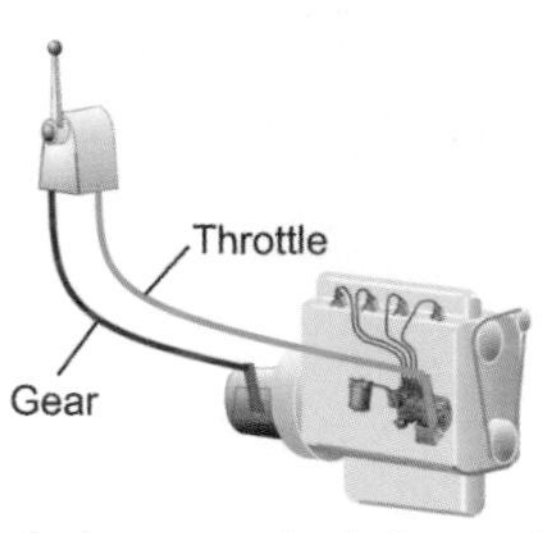

Single lever controls to engine

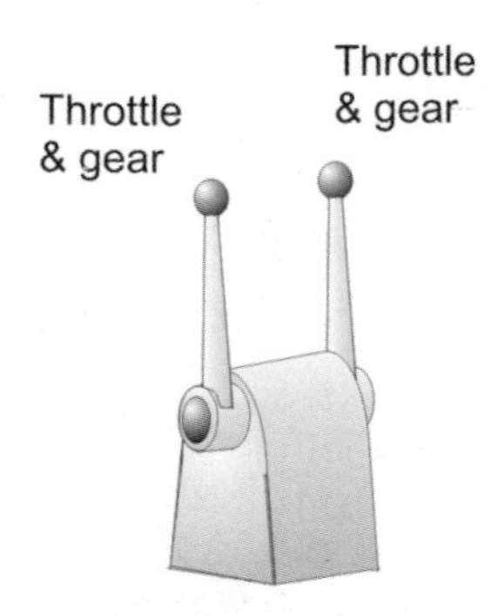

Single lever
– twin engines

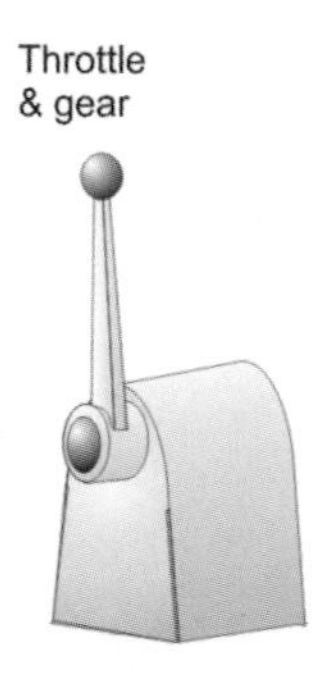

Single lever
– single engine

is adjusted to ensure that the ahead or astern gears are fully selected before the throttle is opened from its idle setting. This system is much more instinctive for the helmsman, but it does introduce extra friction and tends to be quite 'notchy' when selecting ahead and astern.

The mechanism inside the 'throttle box' consists of a number of gears and levers as shown.

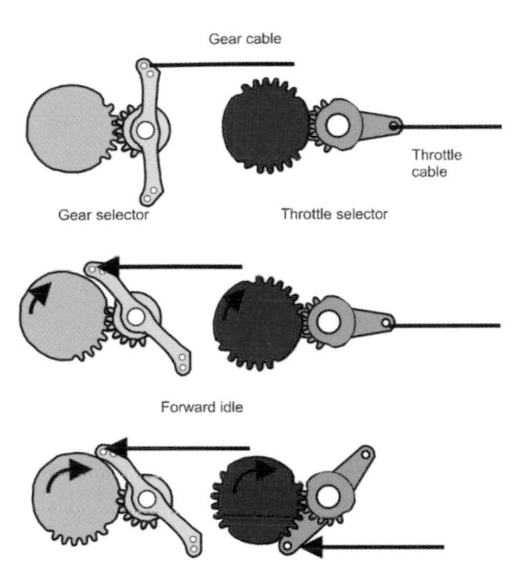

Single lever control – forward

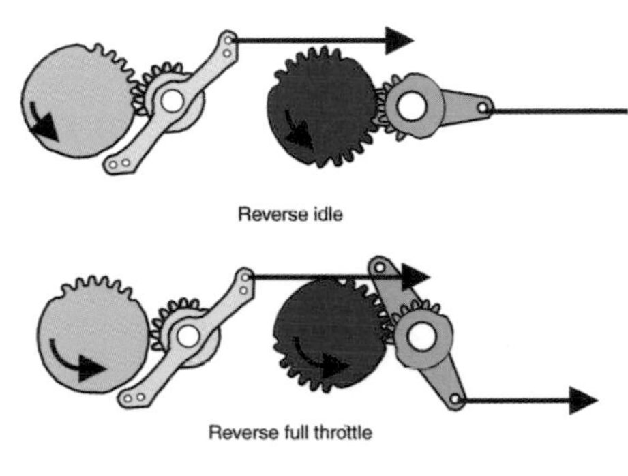

Single lever control – reverse

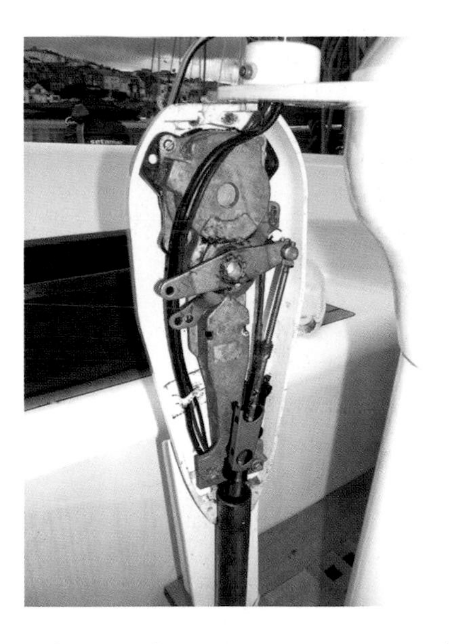

Because it's sometimes necessary to open the throttle but keep the gearbox in neutral, there has to be a means of disengaging the throttle linkage from the gear linkage. This is achieved in one of two ways: either by pushing a button at the centre of the throttle lever boss (or below it) and holding it as you push the lever forwards or backwards, or by pulling the whole throttle lever out away from the housing as you pull or push to increase rpm.

On higher-powered engines it is the practice to prevent the engine from being started in gear. In this case, a 'micro-switch' mounted in the mechanism is used. This 'normally open' switch is closed when the gear selector is engaged in forward or astern. The positive wire from the starter switch to the starter solenoid or relay is run via this switch, so that the starter will not be activated if the selector is not in neutral.

Tip

If the starter will not turn, check that the gear selector is in neutral.

DUAL-STATION SYSTEMS

It's sometimes desirable to have control of the throttle/gearbox from two different positions in the boat. This requires a relay system, so that the engine may be

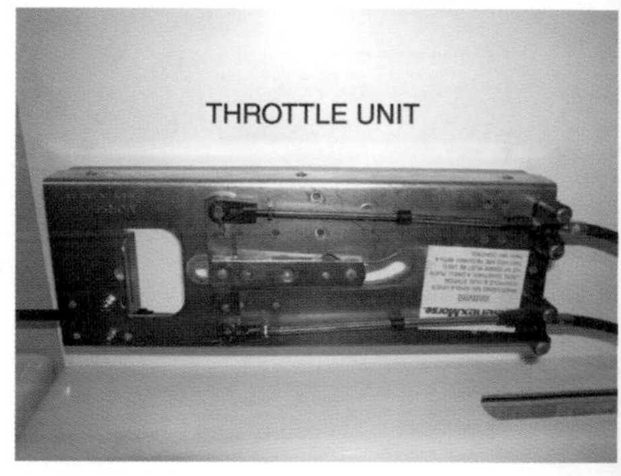

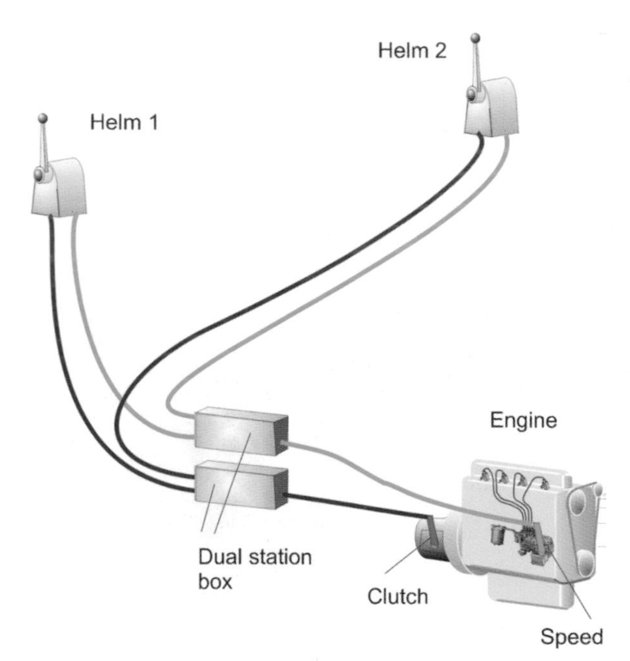

Gear and throttle controls

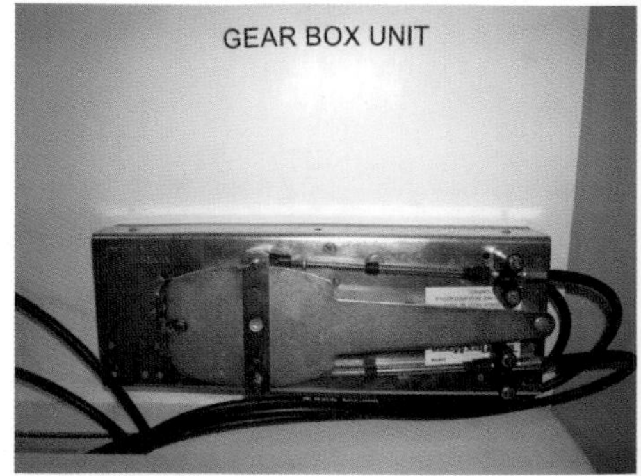

operated from either 'station'. The controls have to be set up properly for correct operation, and the 'feel' of the levers is heavier and more 'clunky'. 'Sympathetic' operation by the helmsman is required for smooth selection of gear and throttle.

The controls will work from only one station at a time, and selection of which station is in control is made either manually or automatically. Both selectors must be in neutral at the time of changeover, which is then achieved by operating a lever in the case of a manual system, or by moving a lever away from neutral in the automatic system.

The operation of the dual station throttle control mechanism is shown in the first diagram. The operation of the dual station gear selector mechanism is shown in the second and third diagrams.

ELECTRONIC SYSTEMS

Modern electronic throttles and gear selectors are very smooth in operation but rely on the electronics working properly.

Electronic throttles

The electronic control system can be overridden, in case of emergency, by following the instructions in the handbook and on the throttle actuator box.

Tip

If the lever of one station is 'stuck', check that the lever at the other station is in neutral and that the manual selector, if fitted, is in the correct position.

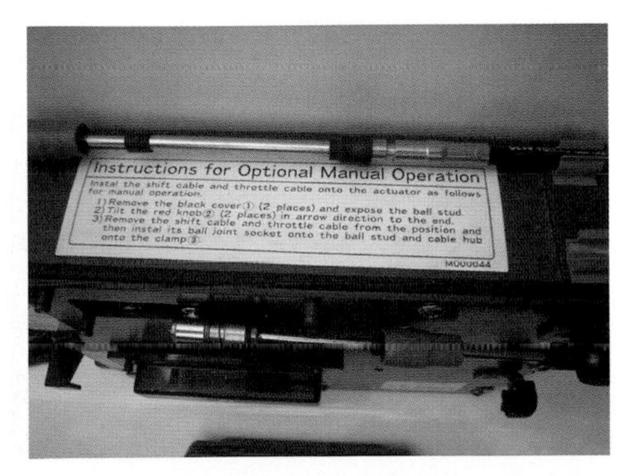

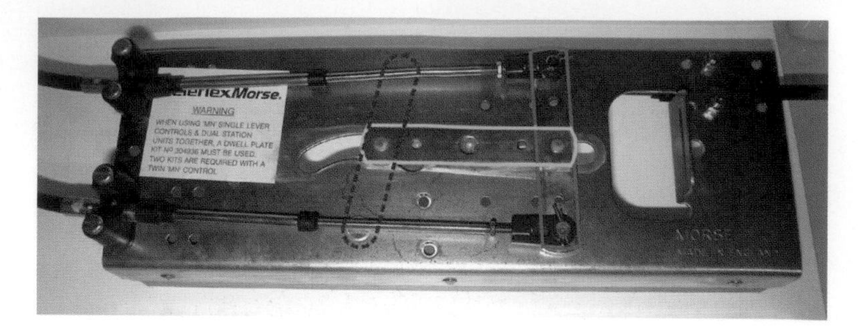

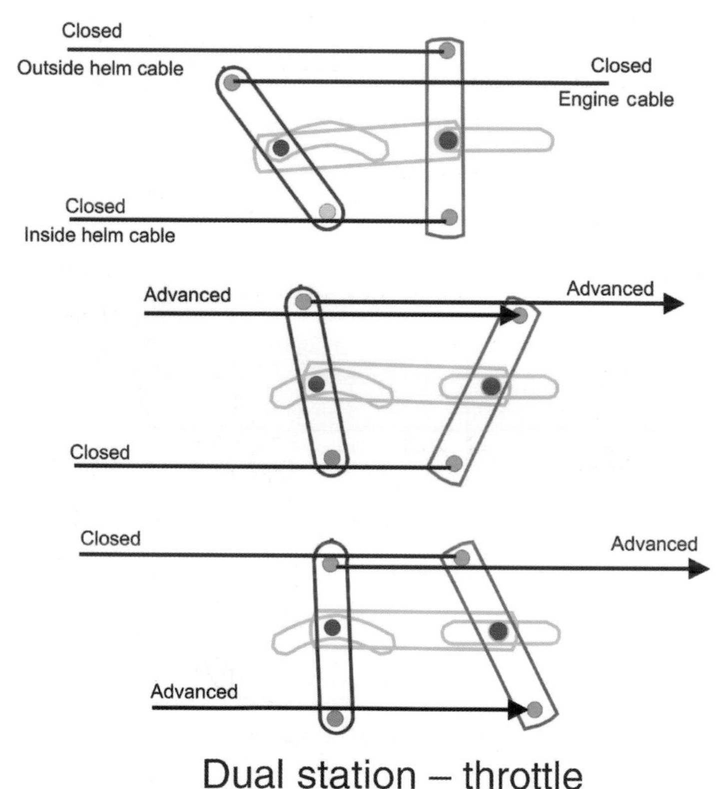

Dual station – throttle

OPERATING CABLES

Operating cables need to be routed so that there are no sharp bends. Bends promote wear and introduce friction. The minimum radius is 100 mm, but aim for greater. Morse makes a special cable capable of radii less than 100 mm, but it's more expensive.

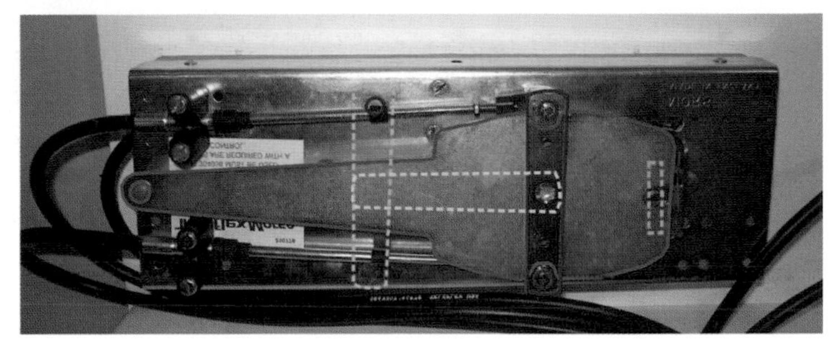

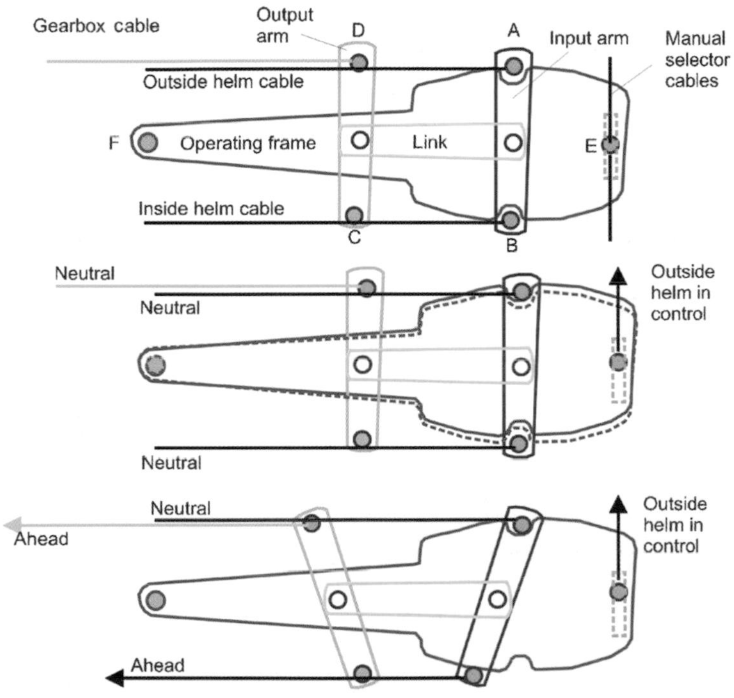

Dual station–gear

Once your cable is more than about five years old, it's worthwhile having spare cables. It would be sensible to route the new cable alongside the old, to reduce the time required to change the cable if necessary. If you need to change the cable quickly, then you may be able to pull the new through with the old as you withdraw it.

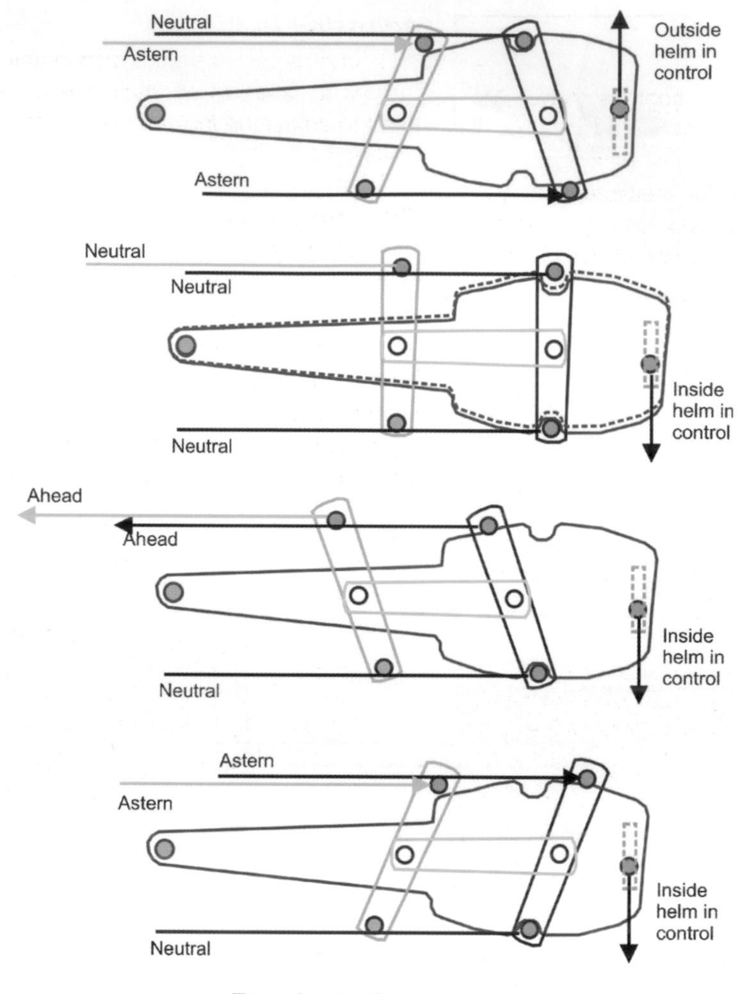

Dual station – gear

MAINTENANCE OF THROTTLE AND GEAR SELECTORS

Lubrication and checking for wear should be carried out annually. This will require some disassembly, so some of the steps outlined below may be necessary. Most boaters won't need to install a throttle unit from scratch, but the mechanism will wear and a cable may need to be changed.

Tip

Older Whitlock controls have the adjuster, but no way of accessing it without removing the unit from the binnacle. By careful measurement, you can drill an access hole so that you can make any adjustment with ease.

Adjusting friction

Most control boxes have a friction adjuster to prevent the throttle closing by itself due to vibration. If you need to adjust the friction, do so with the throttle lever advanced.

Changing a Whitlock throttle/gear selector unit

Typical of a single-lever control is the Whitlock system fitted to many sailing boats. The older type have a parallel splined shaft engaging on an aluminium splined collar attached to the throttle lever with two grub screws, and are not renowned for their security. Later types have a cast stainless steel lever mounted directly on a tapered spline and are much more satisfactory.

If you need to change the unit, here's how to do it:

1. Remove the two grub screws from the base of the throttle unit.

2. Slide off the lever and collar.

3. Remove the two screws at the top of the 'pod'.

4. Rotate the pod to give access to the four clamp screws.

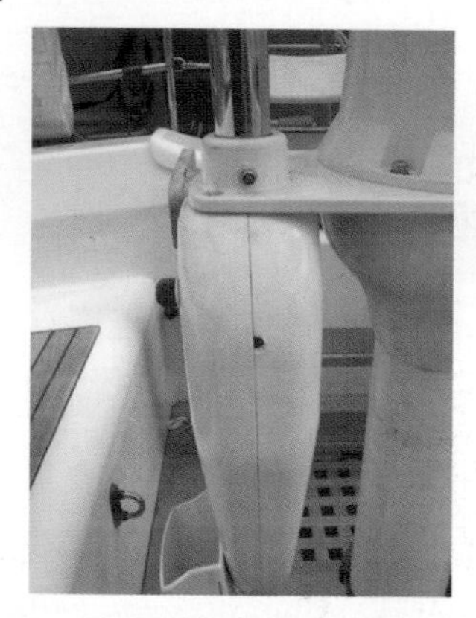

5. Remove the four clamp screws.

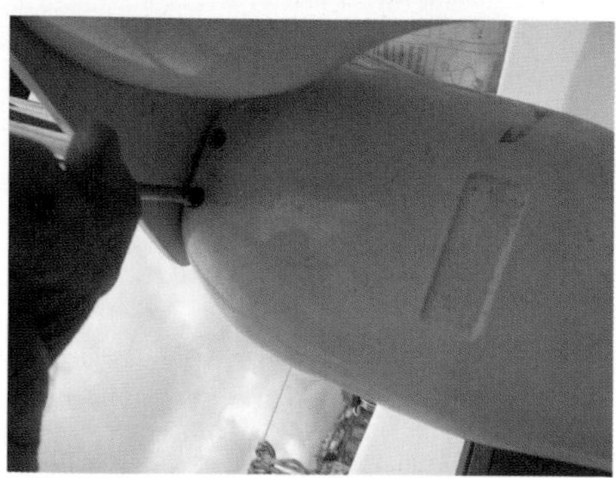

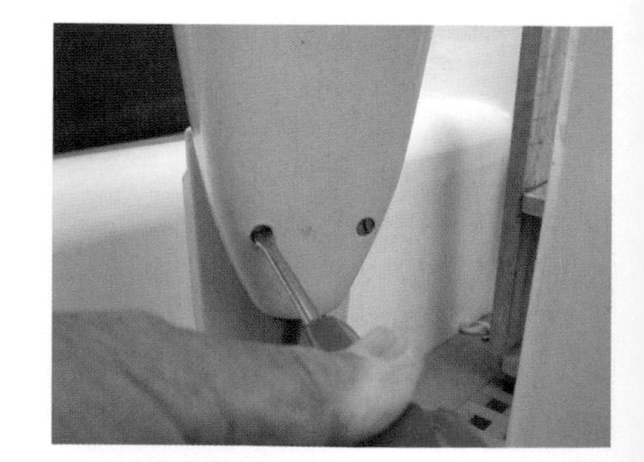

6. Prise the two halves of the pod apart.

7. Note very carefully which cable goes to which pivot point. All holes are numbered, but most are redundant for each specific application. Photographing the unit with a digital camera is a good idea at this point.

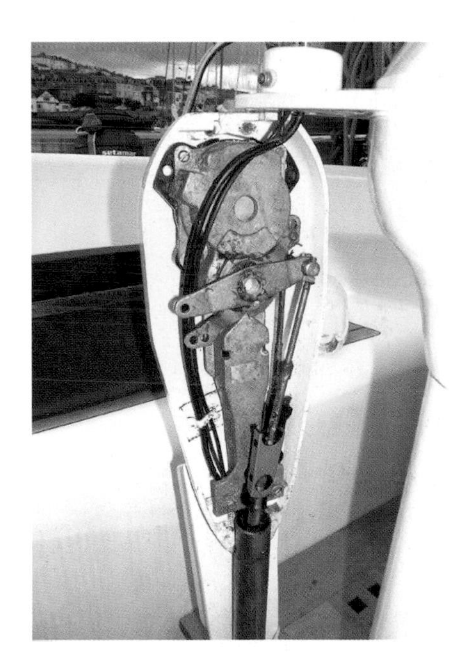

8. Remove the four screws attaching the throttle unit to the pod case.
9. Remove pod case.

10. Remove the split pins from the end of each cable attachment. You may have to operate the lever to gain access to the gear shift attachment.

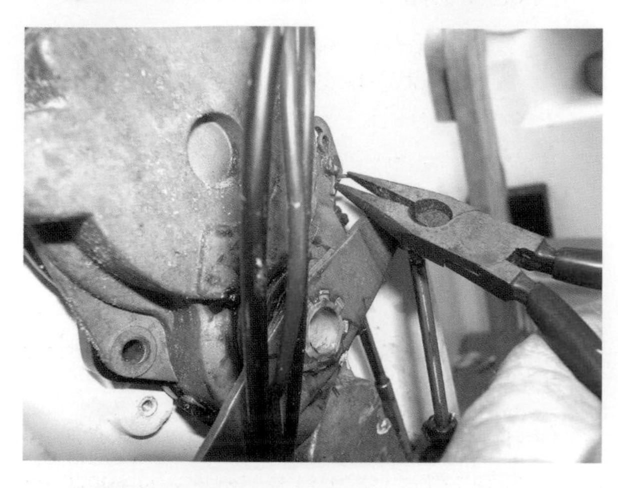

11. Remove the cable snap-on cover from each cable.

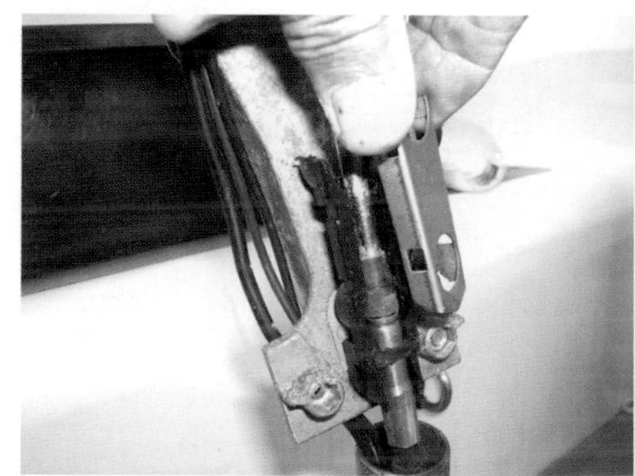

12. Remove the cable clip from each cable.

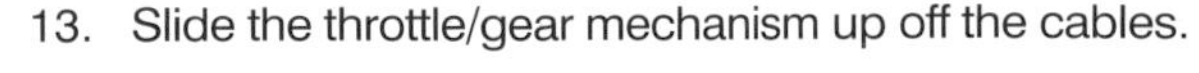

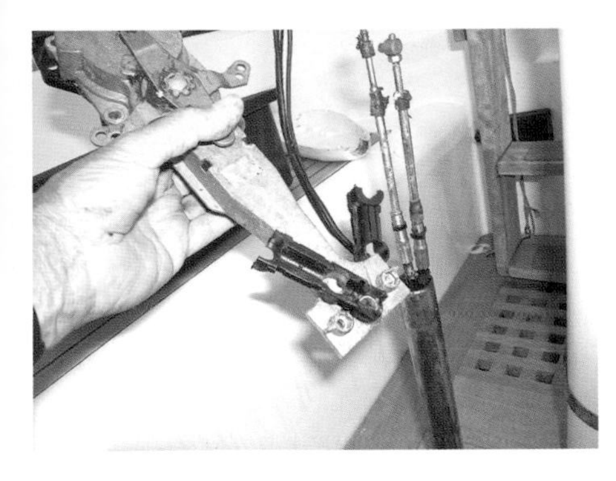

13. Slide the throttle/gear mechanism up off the cables.
14. Disconnect the throttle cable from the speed lever on the engine.
15. Reassemble the new unit in the reverse order, using all the new parts that come with the installation kit.
16. Apply grease to the cable attachment swivels and the cables as you proceed.
17. Reassemble the unit to the pod, checking that the gear selector is in the neutral position before you do so.
18. Attach the throttle lever in the vertical position (or the position of your choice for neutral).

19. Check the operation of the unit.
20. Attach the throttle cable to the engine speed lever, ensuring that it is resting only very lightly against the idle stop. Adjust this end of the cable if necessary.

21. Check the operation of the unit.
22. Insert the neutral button.
23. The rubber cover for the neutral button is glued in place, so don't glue it in yet.

24. Start the engine and warm it up.

25. Pressing the neutral button, open the throttle slowly in both ahead and astern to ensure that you get the full rated maximum rpm to ensure the stops are rigged correctly.

26. Stop the engine, check all screws for tightness.

27. Glue the rubber cover in place with three blobs of the super glue provided.

Replacing throttle or gear selector cables

To replace either or both cables you'll need to follow the steps above up to number 14, then:

1. Route the new cables to replace the old ones, ensuring that any bends have a radius no less than specified by the manufacturer. Some cables have less friction and can tolerate tighter bends than standard cables, but cost more.

2. Reassemble the throttle/gear unit.

3. Reconnect the new cables to the engine and gearbox.

4. Ensure that the speed lever rests lightly on the idle stop and adjust this end of the cable if necessary.

5. Ensure that the gear select lever on the gearbox achieves full travel or the clutches will slip. Reference will have to be made to the gearbox manual for this information.

6. Check full operation as before, both in gear and out of gear.

Electronic Engine Management

Just like modern cars and trucks, control of exhaust emissions is becoming a prime factor in modern marine diesel engine design. A major feature in the control of emissions is electronic engine management, and the latest high-power marine diesels are now equipped with it. As time goes on, electronic management will be fitted to smaller engines as well. The electronic engine management box will be found on the side of the engine.

Again, just like your car, reliability is pretty good, but problems can arise out of the blue, and solving them can be time-consuming and inconclusive if they are intermittent.

WHAT IS ELECTRONIC ENGINE MANAGEMENT?

Electronic engine management controls every step of the operation of the engine and gearbox.

- When you are ready to start the engine, you turn the 'ignition' key to ON. This switches on the electronic

engine management computer, which then carries out a series of checks on itself and the engine systems. If it finds any problems, it sends messages to the display panel and screen and may not permit a start.

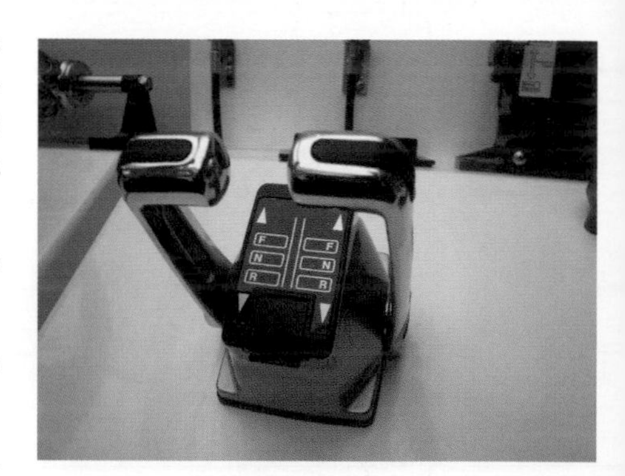

- It also senses the engine and air temperature ready for the start.
- The throttle must be at idle and the gear selector in neutral for the start to be allowed.
- When you press the engine start button, the computer takes any necessary cold start action required and the engine starts.
- Once the engine is running, the computer monitors all the sensors and if problems occur, sends signals to the warning system and the diagnostic display. In extreme cases, it will shut the engine down.
- When you open the throttle, the computer controls the fuel injection in such a way that no smoke is produced, i.e. you can open the lever as fast as you like, but the fuel injected is only as much as can be burned by the air available.
- At all times, the computer matches the fuel injected and the operating conditions of the engine to match the load on the propeller.

Electronic engine control panel

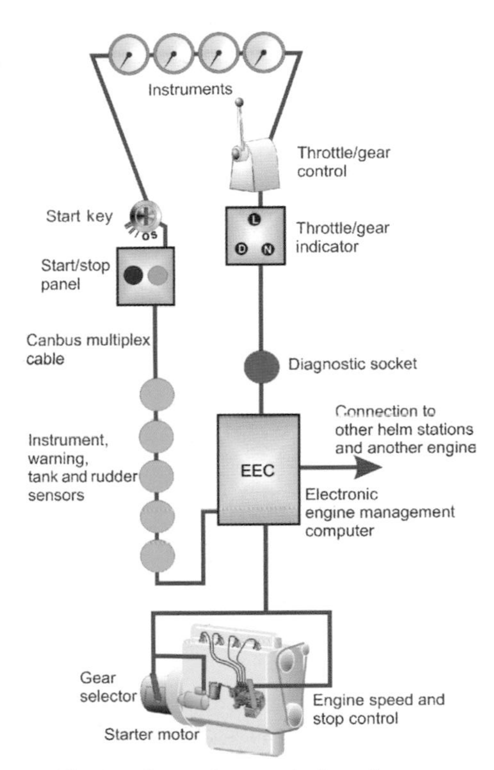

Electronic engine control system

- For certain faults, engine rpm is limited to prevent damage.
- The rpm of twin engines is synchronised.
- On Volvo Penta IPS engines, the computer also controls the steering, limiting the steering angle at higher rpm.
- The engine is stopped using the electronic management system.

All components are wired together with a cable that all signals pass along at the same time (a multiplex cable using the CANBUS protocol). Any component added to the cable can input or access any data it needs.

The engine information can be displayed on digital instruments, as shown in the photograph.

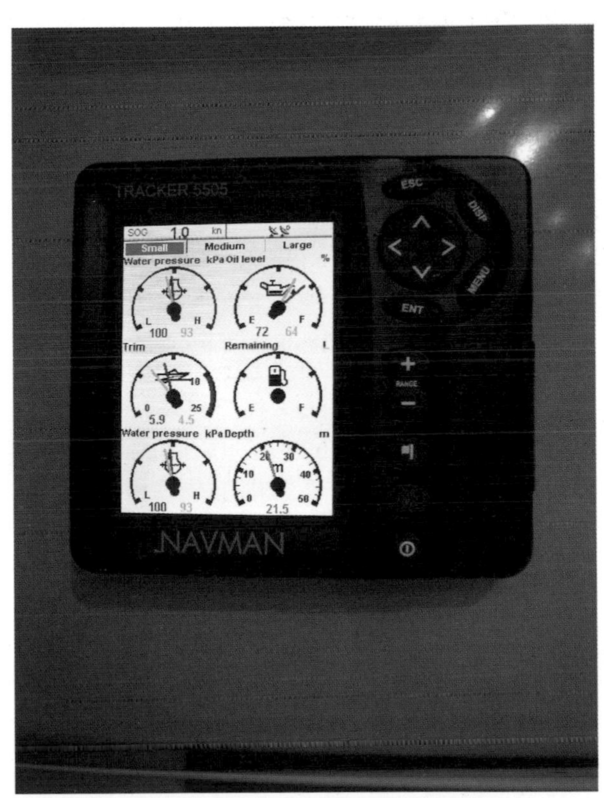

Failure of any part of the system can result in the engine either not starting, not stopping or stopping when you don't want it to; however, like the modern car, as long as all is working properly, operation of the engine requires very little thought.

Because there's no throttle and gear cables or dual-station boxes, throttle and gear selection is friction free, making low speed control very smooth. Electronic throttle and gear selection can be retro-fitted to engines without electronic engine management.

On Volvo IPS installations, a joystick control and additional computer can be added to control engine speed, gear selection and steering input, to make very slow speed control ultra precise, so that the boat can be moved sideways from its berth with a single control!

Routine maintenance by an agent for your engine should ensure that at each service, the service history of the engine can be read from the computer via a laptop or PDA, and any pending problems rectified.

TROUBLESHOOTING

With electronic engine management, troubleshooting becomes both easier and more difficult at the same time. Easier because all you have to do is plug a laptop computer or PDA into the diagnostic socket to find out what's wrong, and more difficult because you have to go to a main dealer to find someone with the equipment and the knowledge of how to interpret the information.

Your engine handbook will tell you what the various indications mean and what to do. It would be an excellent idea to photocopy the relevant diagnostic pages in the engine handbook and have them readily to hand as an aid to troubleshooting.

As far as the owner is concerned, there is little he can do except read the diagnostic panel, get out the handbook and find out what the indications mean. There are few connections that he can check, except to see if 12 volts is available at the starter motor solenoid if the engine won't start.

As with all computers, the first starting point in case of a problem is to switch everything off, wait ten seconds and switch it all back on!

In certain conditions, you may be able to limp home at reduced rpm or even manually select forward gear, but then the only way of stopping the boat is to stop the engine, as you can't select reverse. Full details should be found in the engine operator's handbook.

On Yanmar engines, the emergency stop control is on the start panel with a red guarded switch. On Volvo engines, you'll need to press the emergency stop button on the engine itself.

Propellers

INTRODUCTION

Propeller design is a bit of a black art. There are books dealing with the subject that allow you to calculate, rather laboriously, the shape and size of a propeller for a given application. Now there's computer software to do the job effortlessly. What do they have in common? All the methods are based on trial and error, they use assumptions, and at the end of the day will get you a prop of roughly the correct design that only sea trials will verify. With as much computing power and theory as is available to design aeroplanes, we could get pretty well spot on first time, but in reality, that's never going to be available to the leisure craft industry.

A PROPELLER

Look at your propeller and you'll find stamped on the hub something like 17 × 11 LH. The first number (17) is the *diameter* of the propeller in inches. The second number(11) is the *pitch* of the prop, again in inches. This represents the angle of the blade to the propeller shaft centreline, so why is it in inches and not in degrees? Rotate the propeller on its shaft one turn and it will wind its way forward by the amount of pitch, in this case 11 inches. Just like a screw, after which the prop is named, being driven into a piece of wood. The letters LH tell us that it is has *left-hand rotation*. Propeller rotation is defined when you look at the prop *facing forward* (towards the front of the boat), clockwise being right-handed and anticlockwise being left-handed. So, in this case, our LH prop rotates anticlockwise when viewed from the rear (the gearbox being in ahead). An engine's direction of rotation is defined from the front looking aft, the opposite to the prop!

Don't assume the prop's rotation is in the same direction as the engine's rotation, because the gearbox may rotate the prop shaft in the opposite direction when in ahead. The same engine, when fitted with a different gearbox, can have opposite prop shaft rotation, so assume nothing when specifying a prop.

HOW DOES A PROPELLER WORK?

Let's start by considering what the prop does. It has to exert a force backwards that pushes the boat forwards. Remember Newton's laws of motion? Well, it doesn't matter if you don't, but one of them says that for every force, there's an equal force in the opposite direction. We want to push the boat forward, so we have to push the water backwards.

Well, it's the prop that pushes the water backwards Archimedes possibly invented the screw way back, and Brunel used a 16-foot iron one to push the *SS Great Britain* along. It's a bit like a wood screw as it works its way into the wood. The wood screw pulls itself into the wood with no slip, so that if the thread of the screw moves forward 3 mm each revolution, the screw pulls itself 3 mm into the wood each revolution. But there the similarity ends. The wood screw works with no slip because the wood is solid, which water isn't, unless it's frozen, and then we're going nowhere anyway!

The water screw pulls itself forward only because it can exert a force on the water, and now it's working like an aeroplane's wing. A wing, or keel for that matter, can exert a force only if it's at an angle to the flow. Align the wing to the airflow and all you get is drag, angle it up a bit and you get lift as well. So the propeller blade must be at an angle to the water flow to get any thrust. For this reason, the propeller will

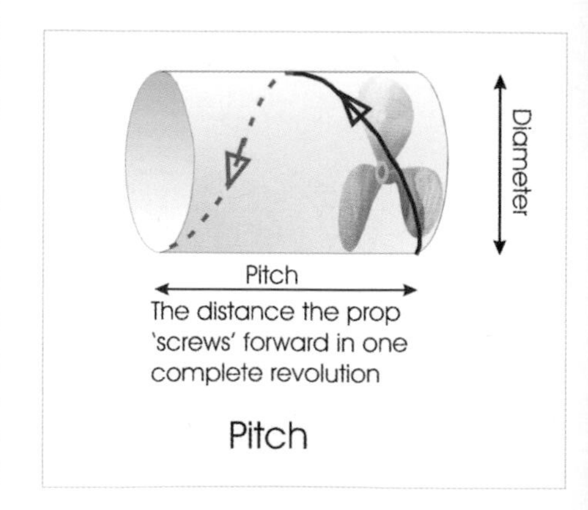

Pitch

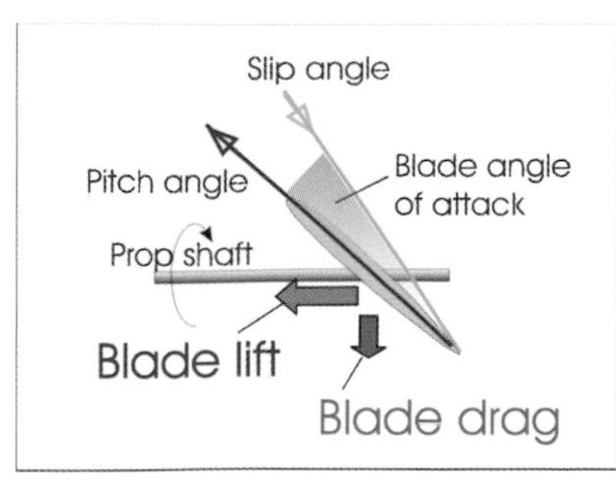

Lift and Drag

always 'slip' as it rotates. Some writers point to slip as inefficiency, but a propeller can't work without slip!

Unfortunately, it doesn't even end here, because to create any thrust, the prop has to be rotating, just like the screw. I know you knew that, but it's here that we get the problem of slip.

SLIP

Slip is the difference between the pitch of the propeller and the actual amount the propeller pulls itself through the water for each revolution of the prop. If there was no slip, there would be NO thrust. In reality, it's the slip that creates thrust, and thus we get propulsion. The amount of slip is determined by a balancing act of water density, blade shape, boat drag, engine power, gearbox ratio and other factors including the hull shape, all ending up in equilibrium.

If the boat is tethered it doesn't move forward, so there's 100% slip. The thrust generated with 100% slip is the *Bollard Pull*, but this has no real relevance unless you are designing a tug boat.

Experience has taught propeller designers that the value of slip appropriate to smaller sailing yachts and displacement motor cruisers varies from around 40% to 55% when at maximum boat speed.

The slip actually experienced in any given situation is related closely to the propeller's efficiency.

Another way of looking at slip is to look at how much water the rotating propeller moves.

- With no slip, no water is moved backwards by the prop.
- With 50% slip, a water volume equivalent to the propeller disc area times half the pitch is moved backwards (i.e. 50% of the water is moved backwards).
- With 100% slip, the whole water volume would be moved backwards.

EFFICIENCY

The efficiency of a propeller is the ratio of the force that is making the prop turn compared to the thrust

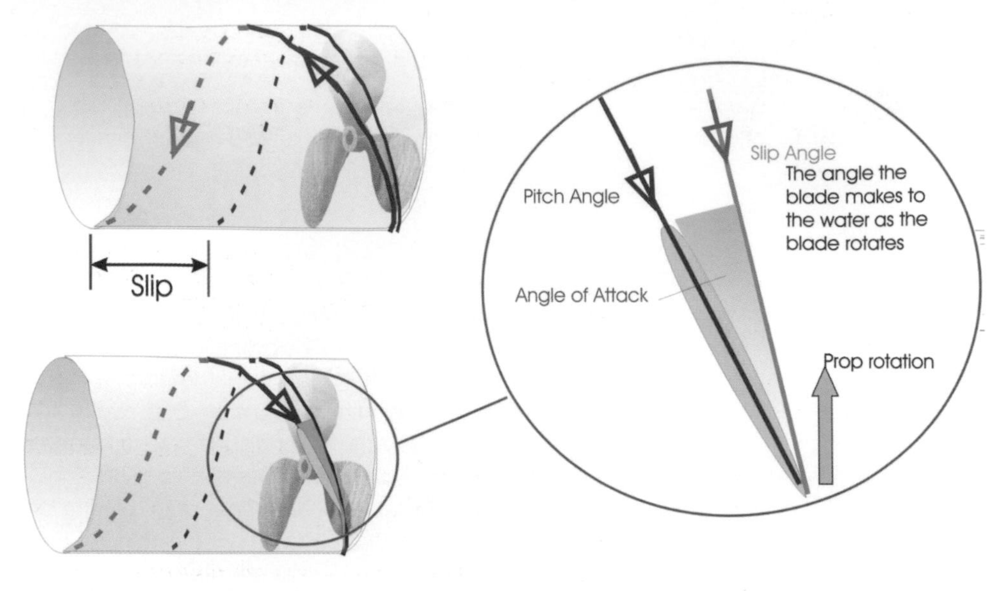

Slip 1

that the prop delivers. Major factors affecting the prop's efficiency are: the plan shape of the blade, the cross-sectional shape of the blade, the ratio of the pitch to the diameter, whether the blade has correct twist and the angle the blade makes to the water entering the prop, which of course depends on the slip.

The blade shape chosen will vary according to the speed and type of hull, and in the case of a sailing boat, whether performance under power or sail is paramount.

THRUST

The thrust developed by the propeller is determined by its diameter, pitch, blade area, shape and its speed of rotation. The speed of rotation is determined by the horsepower delivered to it. At a steady boat speed, the thrust from the prop equals the drag of the boat.

We are unable to calculate the thrust of the propeller accurately, as we cannot calculate its efficiency. As a

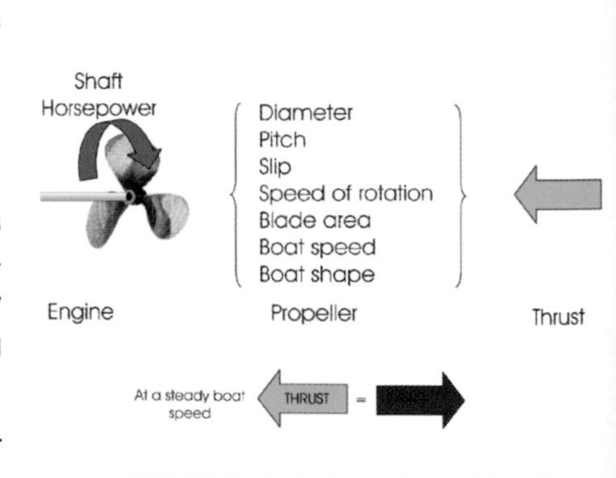

CONVERSION OF ENGINE POWER TO THRUST

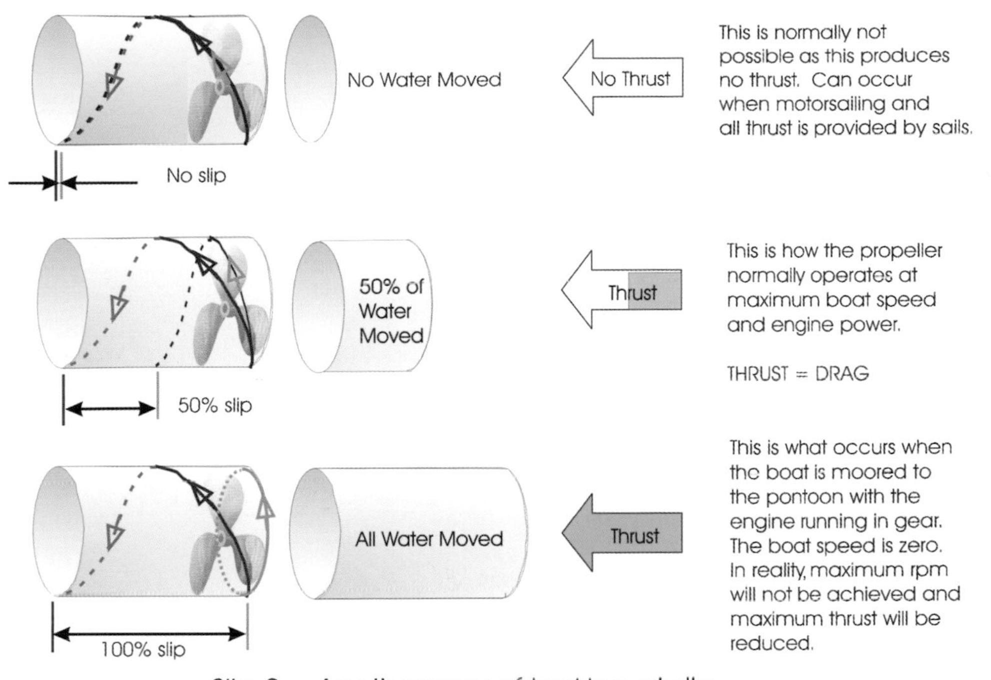

Slip 2 – Another way of looking at slip

result of designers' experience, curves are available relating efficiency, slip and other factors that we are forced to use to calculate the theoretical thrust of a prop. These relationships have been found to vary with the shape of the hull, its speed through the water and the propeller shape.

Diameter

Increasing prop diameter increases thrust, but you need a minimum of 15% of the prop's diameter as clearance between the prop tip and the hull. In general terms, go for the biggest diameter you can.

Pitch

Increasing pitch increases thrust, but only up to the point that the propeller blade stalls. At this point, there's a dramatic fall in thrust and a large increase in blade drag. Prior to reaching this point, the pressure on the 'upper' side of the blade is reduced to such an extent that *cavitation* occurs.

Propeller manufacturers are able to adjust the pitch of a fixed-pitch prop by up to two inches to 'tune' the prop to the boat if it's not quite right.

Cavitation

When the pressure on top of the blade is reduced sufficiently, air dissolved in the water forms bubbles, which then implode, causing noise, some loss of thrust and erosion of the surface of the propeller. Research has established when this is likely to happen, and the blade loading (thrust divided by blade area) at which this occurs should not be exceeded.

Cavitation is not to be confused with *aeration*. Air being drawn down from the water surface because of the prop's proximity to it is often called cavitation, but in fact this effect is properly called aeration and it will cause loss of thrust. Unfortunately, many writers confuse these two entirely different processes.

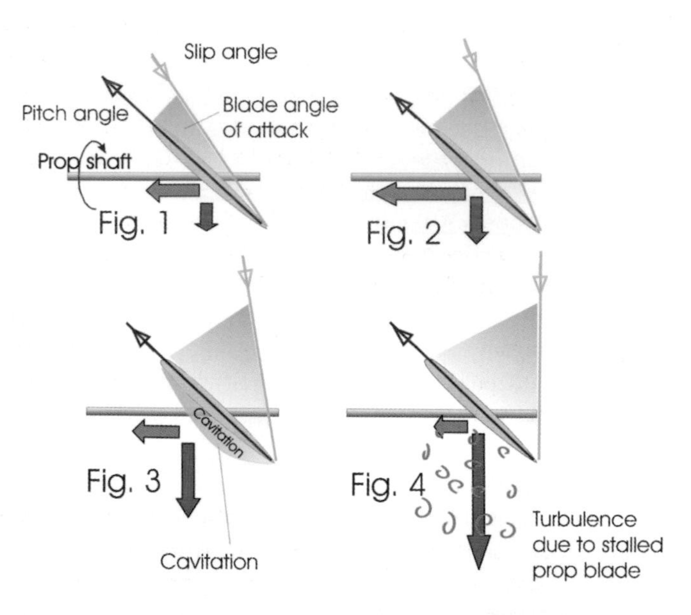

Propeller Thrust

Propeller drag (the greater the prop drag, the more the engine torque needed to turn it)

The best operating range for a propeller will be found with a pitch somewhere between figure 1 and figure 2, where there's the best compromise between the thrust produced by the prop and the torque (power) required to turn it.

Cavitation

BLADE AREA RATIO = AREA OF BLADES COMPARED TO AREA OF FULL CIRCLE

20%
Typical two-blade folding propeller

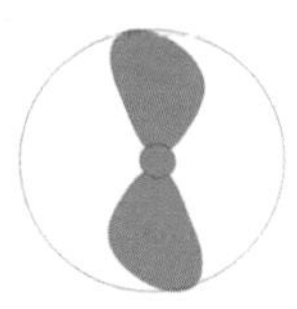

30%
Typical two-blade fixed propeller

45%
Typical three-blade sailing propeller

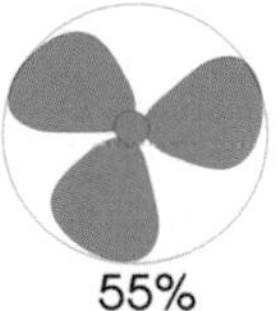

55%
Typical three-blade displacement motor boat propeller

For a given angle of attack, a larger blade area will give a greater thrust

100% BLADE AREA RATIO

Blade area

Up to the point at which blade loading makes cavitation likely, the blade area is not a big factor, as reduced blade area is made up by increased pitch. If the prop is working close to the cavitation point, extra blade area will reduce the likelihood of cavitation.

Blade area is normally considered as the blade area ratio: the area of the blades divided by the area of the blade diameter circle.

So, if the prop is working well within its limits, such as when cruising, a two-bladed prop (of correct pitch) will produce a thrust similar to a three-bladed one. In fact, because each blade has more room, it is likely to be more efficient than a three-bladed prop and deliver more thrust. However, when the two-bladed prop is running under adverse conditions, such as rough weather, or under maximum acceleration, the three-bladed prop will then deliver more thrust, because its blade loading will be lower.

This can be seen in the table, where the thrust and maximum speeds of all three types of 17 inch diameter props are very similar, although the performance of the two-blade folding prop (which has narrow blades of small area) has started to fall due to overloading. However, only at lower blade loadings is there any margin of performance to allow for adverse conditions. The margin of the folding two-blader would be improved by changing the gear ratio so that it turns more slowly and then increasing its diameter to 18 inches. The table shows some prop calculations for a twenty horsepower engine driving a two-ton boat at maximum speed.

Blade twist

The speed at which the propeller blade meets the water varies along its length, because the tip speed of rotation is much faster than at the root. In order that the blade's angle of attack is the same all the way along, the propeller blade must be twisted.

Table 1						
Gear ratio	Prop	Type	Max speed	Thrust	Blade	Notes
	D x P		kts	kg	loading%	
2.5:1	17 x 13	2-blade folding	7.5	191	104	overloaded
3:1	18 x 16	2-blade folding	7.7	202	94	OK
3:1	17 x 15	2-blade	7.7	203	80	OK
3:1	17 x 12	3-blade	7.6	202	52	Good

Blade loading of greater than 100% is overloaded, causing cavitation.
Blade loading of 90 to 100% is close to overload and any adverse conditions will cause overloading to occur.
Blade loading of 80 to 90% has some margin for adverse conditions.
Blade loading of less than 80% has plenty of margin for adverse conditions.

FIXED PITCH PROPELLERS

A fixed pitch propeller is the correct size only for one set of conditions. Normally, it's designed to be correct at the boat's designed maximum speed and with the engine developing maximum power at the rated (maximum) rpm. Under any other conditions it's the wrong size! This is because the angle at which the blade is meeting the water changes as the forward speed of the boat changes.

WATER FLOW PAST THE PROPELLER

We might well expect that the water flow into the propeller disc is the same as the boat's forward speed, but life's never that simple. Because of the friction between the hull and the water, the flow close to the hull is slowed down, and the water's speed at the rear of the hull is reduced. Also, because the hull is curving in towards the stern, this exerts another slowing effect, and if the prop is mounted in a cut-out in the keel, the slowing effect is exaggerated. Depending on the hull's shape and size and how close the prop is to the hull, this reduction of water speed into the propeller may be as much as 40%.

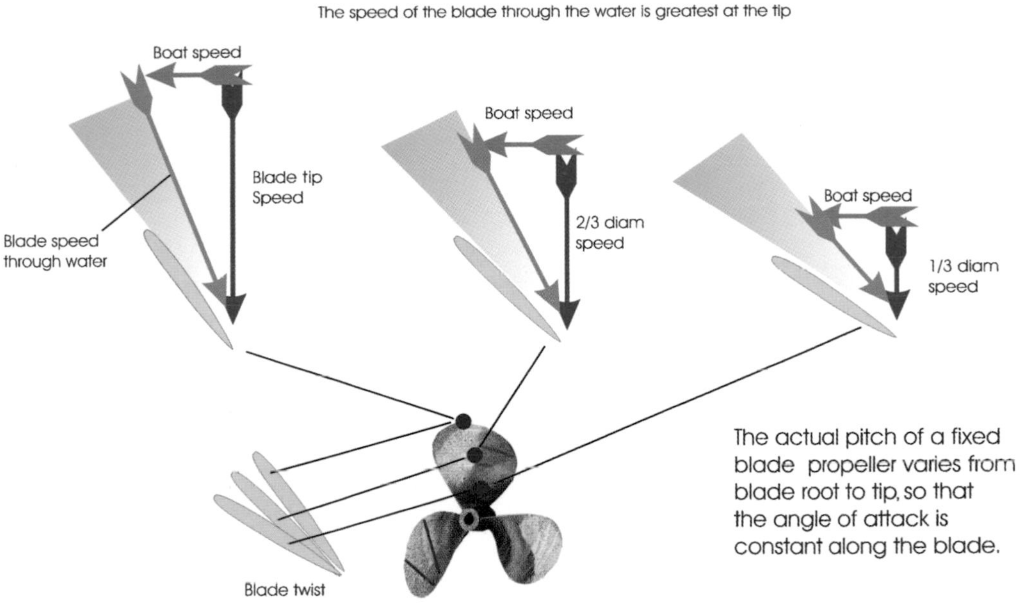

The speed of the blade through the water is greatest at the tip

Boat speed
Blade tip Speed
Blade speed through water
Boat speed
2/3 diam speed
Boat speed
1/3 diam speed
Blade twist

The actual pitch of a fixed blade propeller varies from blade root to tip, so that the angle of attack is constant along the blade.

BLADE TWIST–Same angle of attack all along the propeller blade

So, when calculating the propeller size, we must use only a proportion of the boat speed in order to get the flow into the prop. Like slip, there are no hard and fast figures, so again, experience supplies the numbers. Various authors use figures between 20% and 40% as the reduction in water speed into the prop for displacement motor boats and sailing boats. The more the prop protrudes into the fast-flowing water stream, the lower the number.

DOING THE SUMS

There are a number of books available on the subject of propeller calculations. Most propeller manufacturers have their own software to do the job, and doing it this way is far easier than wading through lots of graphs. Even so, the software is based on graphical solutions incorporating approximations of the various factors. In the end, the calculations will get you into the correct 'ball park', but fine tuning may be needed after sea trials.

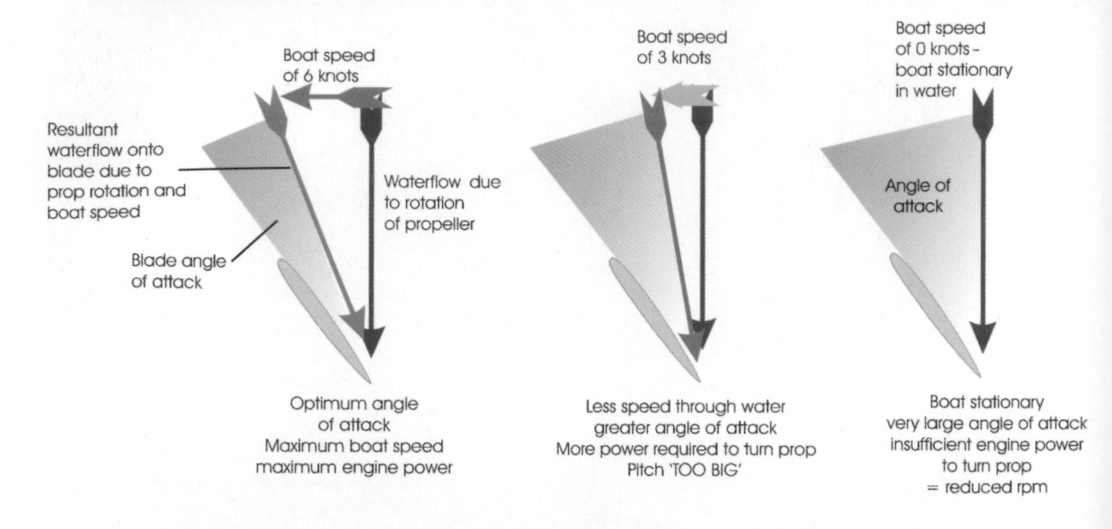

PROPELLER SIZED FOR MAX SPEED, POWER & RPM

I believe that it's best to let the engine/propeller supplier carry out the calculations, then if the prop is not correct, at least there's some comeback.

PROPELLER OVERVIEW

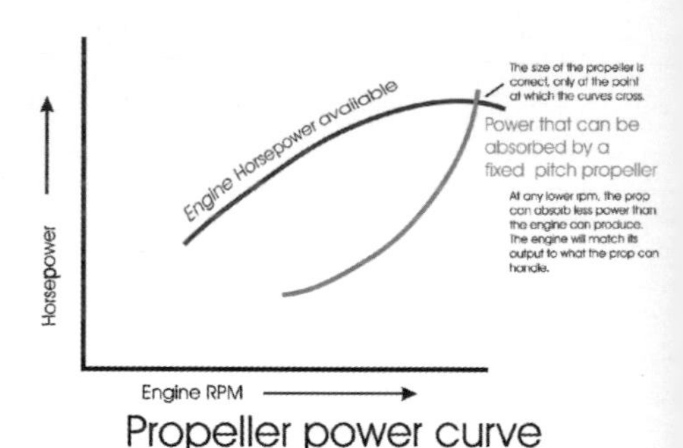

Propeller power curve

There's always a conflict between performance under sail and performance under power. It was ever thus. But even for a displacement motor boat there's a conflict. The prop will be the correct size only at one speed. Normally, the prop's sized for maximum speed, so at cruising power, the prop will have too fine a pitch.

Size the prop for cruising revs and the pitch will be too great at higher rpm, the engine won't deliver maximum power and the boat won't reach its design speed. Many owners, given the choice, will opt for the 'cruising prop' and accept a lower top speed. This gives more relaxed cruising and better miles per gallon. In these circumstances, there's likely to be dark smoke in the exhaust at maximum throttle, and that is frowned on by our friends in Brussels. They would prefer us to

Full-bodied hull with full-length keel and propeller set in an aperture – water speed into prop reduced by as much as 40% in this case.

Well-exposed propeller with a long-fin keel and a moderate hull – water speed into hull reduced by about 20 to 30%

Water flow into propeller

be 'under-propped', eliminating the smoke at full revs but restricting thrust and hence top speed. Revs would need to be higher at our cruising speed, making life less peaceful.

When it comes to sailing, we would like our boat to have minimum drag. For this we need a small blade area propeller or, even better, a folding or feathering one.

How much room does the prop need?

Often, the propeller size is constrained by the clearance around it. There should be a minimum of 15% of the blade diameter clearance between the blade tip and the hull. Often that's all there is to it. However, if the prop is mounted in a keel cut-out, prop size may become more of a problem. Once you try and move away from a conventional fixed prop, you may need room for the blades to fold or swing, and lack of room may dictate that only a certain type of prop will fit.

If you fit too big an engine, and prop size is restricted, you may not be able to utilise all the engine's power.

You will not be able to utilise the engine's full rpm because the prop will cavitate. At 100% blade loading, where cavitation is likely to occur, you will be restricted to 16 hp, but at a more reasonable blade loading, rpm should be restricted so that only about 13 or 14 hp is generated. All in all, under these circumstances, virtually all the increased power available has been wasted!

Power	rpm	Gear ratio	Diameter	Loading
New engine 20	3600	2.5	16	117
Old engine 12	3000	2.5	16	83

Propeller restricted to 16 inch diameter. Two-blade folding prop, blade loading restricted.

Fixed propellers

A fixed blade prop has very good propulsive efficiency at the speed for which it's designed when under power. In astern, the leading edge becomes the trailing edge, because the direction of rotation is reversed and thrust is reduced by as much as 50%. Under sail, the drag is very high and for a three-blade prop a reduction in speed of 15% (0.9 kts at 6 kts) is possible when compared with a feathering prop. Typically, a fixed three-blade prop will have a blade area ratio (BAR) of about 50%, i.e. the area of the blades will be 50% of the area of the circle defined by the diameter. A two-blade prop will have a BAR of about 33% and so will have less sailing drag than the three-blader. Slim two-blade props of 25% BAR are also available.

Propellers for planing motor boats

There's even more difference between maximum speed requirements and low speed requirements for high-speed craft. Obviously, the propeller will be designed for maximum speed and the prop will be inefficient at low speed. A particular problem occurs at the speed at which the boat just starts to plane. A lot of thrust is needed to get over the 'hump', but the pitch is completely wrong for this speed. Specialist high-speed props have been developed, and one gearbox manufacturer has a two-speed gearbox so that you have a gear ratio matched for low-speed work up to just above the 'hump', and a higher gear for high-speed cruising. If the engine power available is marginal, an incorrect prop can even prevent the boat getting on the plane.

Folding propellers

Folding props have considerably less drag than fixed props when sailing, but some tend to slam or jerk as they open and there's some lag as the throttle is opened. They tend to be pretty poor in astern, because they don't open fully, and they may also have less than the optimum amount of twist, and hence they're not so efficient. Some are more complex than others and this is reflected in their price.

- Folding two-blade props have a small blade area (around 25% BAR), so end up with high blade loading and a risk of cavitation.
- Folding three-blade props have more blade area, so run at a satisfactory blade loading but still have little twist.
- They need room behind them to allow the blades to fold backwards.

Feathering propellers

With the engine switched off, the blades of feathering propellers turn at their roots instead of folding, so that the blades are edge on to the water flow past the prop and have the lowest sailing drag of all propellers.

They come in two- and three-blade forms and have a greater blade area (similar to a fixed prop) than their folding brethren, so have a reasonable blade loading and are less susceptible to cavitation. However, they have no twist at all, so lose out on propulsive efficiency, but are much better than a fixed prop in astern, as the leading edge of the blade remains the leading edge – the blade flips round when going backwards.

These need room for the blades to swivel, but don't need a lot more room than a conventional prop.

Self-pitching propellers

To overcome the problem of there being only one boat speed/engine speed combination that is correct for the prop, we could arrange for the propeller to have variable pitch, and then we could have the best of all worlds. This is done on many propeller-driven aircraft, but it's costly and needs two levers – a power lever and a pitch lever.

Brunton Propellers have developed their Autoprop, which cleverly balances centrifugal force, water load on the blade and engine power using swivelling blades, to give a continuously variable pitch matched to the load. This means that over quite a large boat speed/engine power band, the prop is always the 'right size', with the claimed advantages of better fuel efficiency, more thrust and lower cruising rpm. It also works well when motoring in rough weather.

Sailing drag is low, only a little more than a feathering prop, and, like the feathering prop, the blades flip round in astern, so giving good reverse thrust. Additionally, the blades are twisted properly to maintain a constant angle of attack. Annual maintenance is required to lubricate the ball races, and it would be as well to invest in the special prop puller to avoid damage if you have to remove the prop.

They need room both in front of and behind the prop to allow the blades to swing.

Variable pitch propeller

RPM	HP	PROP D x P	SLIP %	BLADE LOADING	SPEED knots calculated	SPEED fixed pitch measured
3600	27	17 x 11	42	71	7.3	
3400	24	17 x 11	40	66	7.1	7.1
3200	22.5	17 x 12	42	63	7	6.9
3000	21	17 x 13	44	60	6.9	6.7
2800	20	17 x 14	45	58	6.8	6.3
2600	18	17 x 15	46	53	6.6	6.1
2400	16.5	17 x 16	46	50	6.5	5.6
2200	15	17 x 17	46	47	6.4	5.2
2000	12	17 x 18	46	41	6.1	4.7

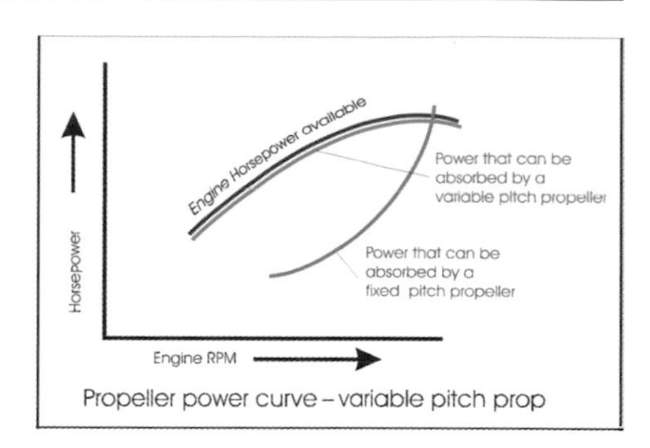

Propeller power curve – variable pitch prop

Note how pitch increases as rpm is decreased. This allows the prop to be correct for much of the engine's speed range, with a corresponding increase in the boat speed for any given RPM less than the maximum as can be seen when comparing the yellow and green columns.

Variable pitch propeller

A standard three-blade fixed propeller gives good thrust but much drag under sail.

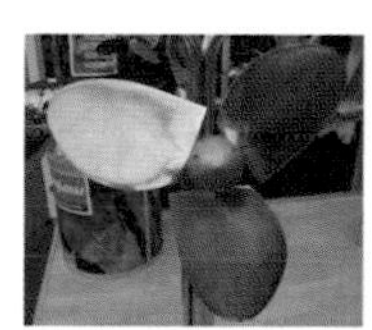

A three-blade feathering prop gives minimum drag when feathered but has no blade twist

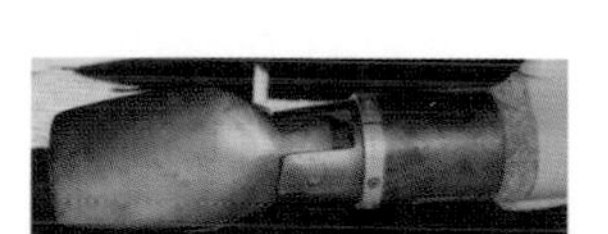

A folding two-blade propeller These and three-blade folding props (below) have proper blade twist but give poor reverse thrust

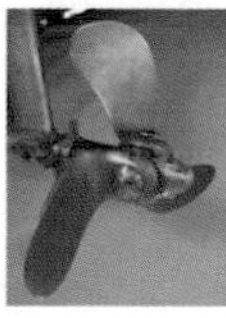

Brunton's self-pitching 'Autoprop' has proper blade twist and adjusts its pitch to the thrust being transmitted

Different types of propeller

Gear ratio

The choice of gear ratio is an integral part of choosing the correct propeller, especially where the diameter or the blade is restricted. This fact often seems to be ignored when installing a replacement engine, especially if engine power is being increased.

Power hp	rpm	Gear ratio	Diameter inches	Loading %
20	3600	2:1	13.9	75
20	3600	2.5:1	15.8	60
20	3600	3:1	17.5	50

Anodes

The more metal there is in the prop, the more anodic protection it needs. Some folding and feathering props have anodes incorporated, but these are relatively small and may not last a full season. Newer Volvo Penta sail drive legs are electrically isolated from the engine and cannot be protected by any hull anode – they rely on the anode at the bottom of the leg. In my experience, their heavy three-blade folding prop does exhibit signs of surface corrosion, so timely replacement of the leg anode is important.

How to make your choice of prop type

The sailing yacht owner has to decide the order of importance of the various factors. Is speed under sail paramount? Is cost all important? Is thrust under power the primary concern?

For a 35-foot sailing boat, powered by a 30 hp engine, the following gives some idea of the effect of the various choices.

It's important to read the propeller performance curves with a certain amount of caution. What you

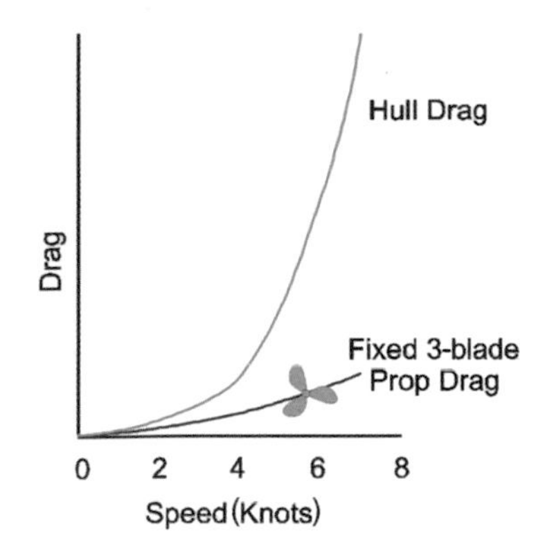

get with a given prop depends on what boat it's used on. The same prop will give a different performance on two different boats, so I have taken all the 'numbers' out of the performance figures, which have been derived from a number of different sources. The curves give an indication of what the different props are capable of to give you a starting point.

At low speeds, the sailing drag of a three blade fixed prop is very significant, but reduces at higher hull speeds, when wave drag of the hull takes effect. You can estimate the increase in sailing speed that can be attained by removing the drag of the prop. You can see that the improvement is best at lower speeds.

The sailing drag of various types of prop is shown. The fixed three-blade prop has significant drag, followed by the fixed two-blade prop. All the feathering and folding props have significantly reduced drag.

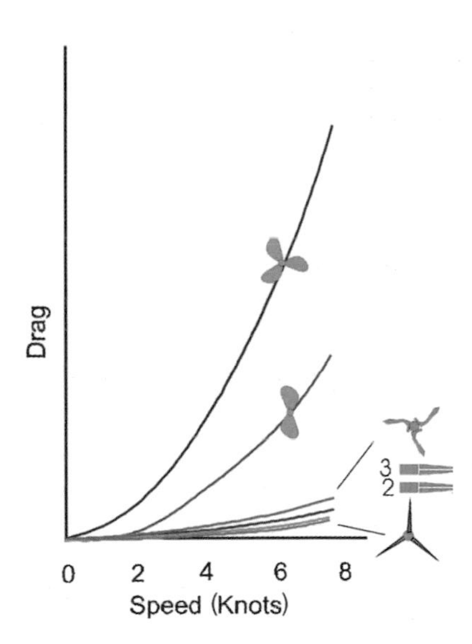

Forward thrust of all the props except the Brunton Autoprop (self-pitching) follows the propeller law principle of being matched to the engine power only at the 'design' speed.

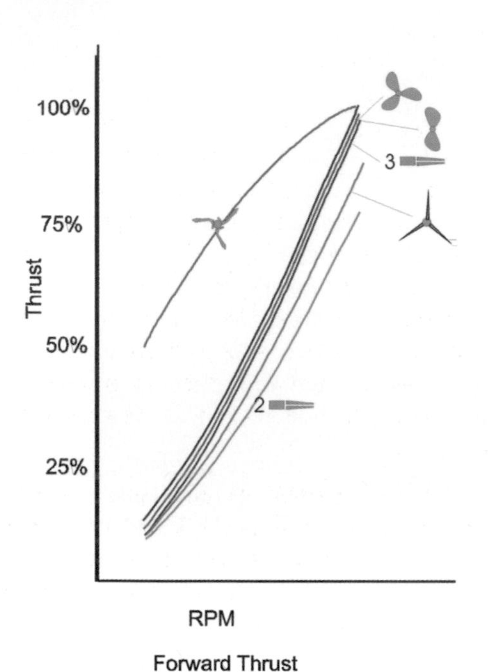

Forward Thrust

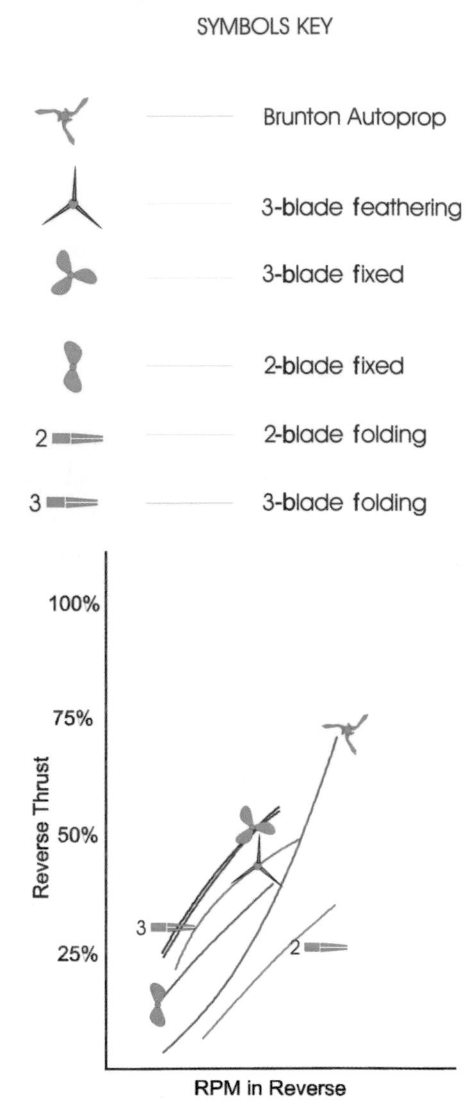

Reverse thrust at 3 knots forward speed

Normally you don't go rushing backwards at 6 knots, so I haven't shown maximum power in reverse. To indicate the stopping power of the prop, I show performance of the prop in reverse with the boat moving forwards at 3 knots and the rpm increased, as you would when approaching a dock. The two-blade folder is particularly poor, and the Brunton, although poor at very low rpm, outperforms the others as rpm is increased, as it adjusts its pitch.

CONCLUSION

Compromise rules all things nautical, and propeller selection is no different. Racing sailors are easy to satisfy because their requirement is minimum drag, but even here there's a choice to be made between an expensive feathering prop and a cheaper folding one.

The displacement motor cruiser's choice is relatively easy, as he wants maximum efficiency at the normal cruising speed. A fixed pitch three-bladed prop will be the normal choice.

It's the cruising yachtsman who has the most difficult choice. My experience of long summer cruises over the last few years shows that I spend probably 50% of my time under power, because we tend to day-sail between ports and 12 hours is long enough. We may not be completely typical, but on passage below about 2 knots, on goes the engine. Also, if the going is heavy and we are going to windward, we often motor-sail at about 20 degrees off the wind, and propeller thrust is important. When the boat was new, we opted for a three-blade fixed prop. The downside of this is that we lose at least half a knot when sailing at 5½ knots, and big waves still reduce our speed, despite the three blades.

I now use a Brunton Autoprop to give better speed both under sail and under power in heavy weather and reduced cruising rpm with no loss of speed. The downside of this is the cost, but I'm delighted with its performance. It has eliminated prop-walk as well.

If you want to have a look at a book of propeller theory that is not too difficult to understand, then have a look at: *Propeller Handbook* by Dave Gerr (ISBN 0-7136-5751-0, published by Nautical Books).

Prevention of Faults

Preventing faults, especially at sea, is very much in the yachtsman's own hands. How can you prevent your engine breaking down?

You can prevent most faults occurring by:

- carrying out regular routine maintenance;
- carrying out your daily checks;
- reading your engine handbook;
- keeping your engine clean;
- understanding how your engine works;
- listening and looking for early signs of future trouble.

Your boat builder can help prevent faults occurring by:

- making a proper installation;
- making the engine and its service points accessible.

In a way, you have control of what the boat builder does, by not buying a boat with a poor installation or poor accessibility. I've been told by prospective buyers that they aren't interested in engine accessibility, as they will be using an engineer to do the maintenance. If you can't get at the water pump or take the engine rocker cover off without using a saw on the structure, neither can the engineer! Installations like this, and I didn't make them up, will, at sometime, result in a lifeboat call out.

The photograph shows the lack of access to remove the sail drive leg. Despite clear directions from Volvo Penta, insufficient clearance above and behind this installation prevents its removal without cutting away significant parts of the athwart-ship and longitudinal bulkheads. Apart from the fact that it isn't possible to remove the leg or gearbox for repair, it's not even possible to change the hull waterproof diaphragm, needed every seven years, unless you take a saw to the structure.

On some installations, the raw water pump is just not accessible! *Caveat emptor*!

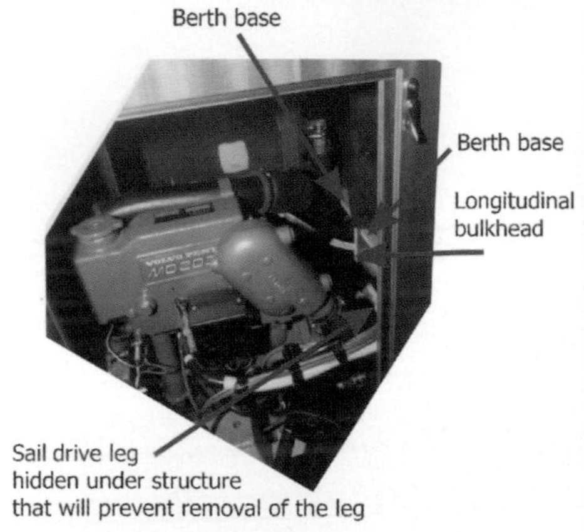

Poor installation

Troubleshooting

Troubleshooting needs knowledge of the system and a methodical approach. The knowledge is contained in this book and the methodical approach is outlined in this chapter.

ENGINE WON'T START

Engine won't turn

- Is the battery 'on'?
- Is the panel alive? – Warning lights 'on'.

  Check the engine fuse – usually on the engine – but there may not be one. Small Yanmars have one but it isn't mentioned in the handbook – it's 30 amp and in the harness close to the starter motor. It's taped over and sprayed with engine paint.

- Check the gear selector is in neutral. Even if there's no neutral switch, the starter won't be able to turn the engine over in gear unless the boat is under way.
- If the warning lights go out when you operate the starter, the battery is flat or the connections very poor.
- If the panel is normal when you operate the starter, then the problem may well be in the wiring or connections to the starter.

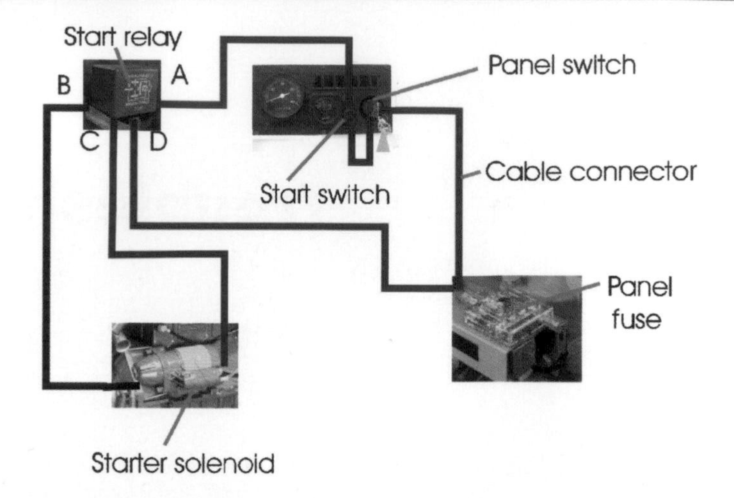

D - B (or any earthing point)		12 VOLTS (or 24 volts)
OPERATE STARTER SWITCH	A - B	12 VOLTS (or 24 volts)
(You should hear or feel a 'clunk')	C - D	ZERO OHMS

- Use the troubleshooting wiring diagram to find and follow the circuit through to the starter motor. Ideally, you will already have tailored this diagram to your boat and identified all the components. There will be various connectors in the wiring loom, some unreliable bullet connectors and others multiple plugs and sockets, which tend to be more reliable. However, unplugging them and then reconnecting may be all that's required.
- With the 'ignition' ON, all the wires shown RED in the troubleshooting wiring diagram should have 12+ volts. Using a multimeter set on DC, volts should show in excess of 12.6 volts between any terminal and a suitable earth terminal.
- With the starter button pressed or the ignition switch held at start, the additional wires shown RED on the troubleshooting wiring diagram should also read in excess of 12.6 volts on the multimeter.

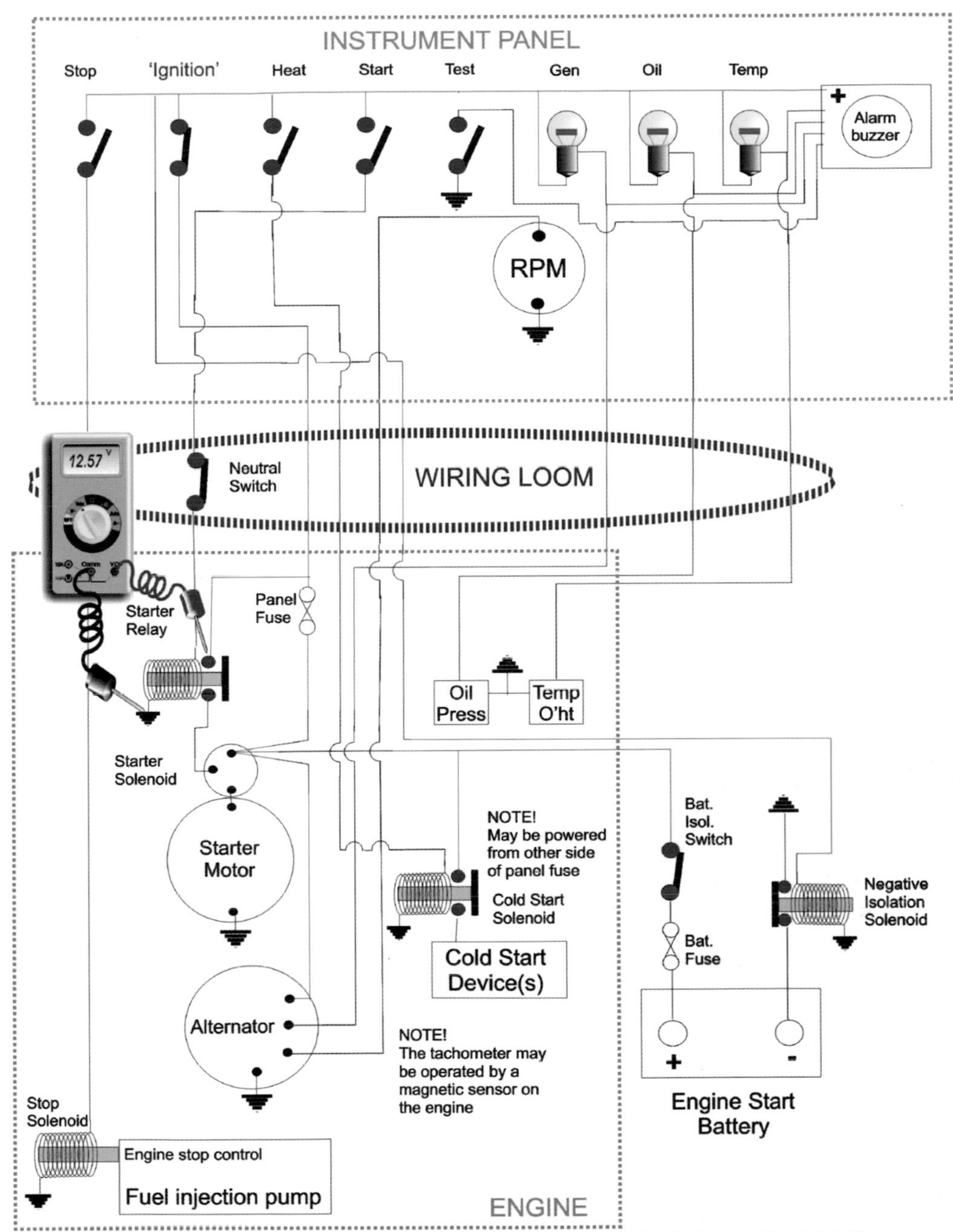

ENGINE ELECTRICS TROUBLESHOOTING
Battery switch and 'Ignition' switch 'ON'

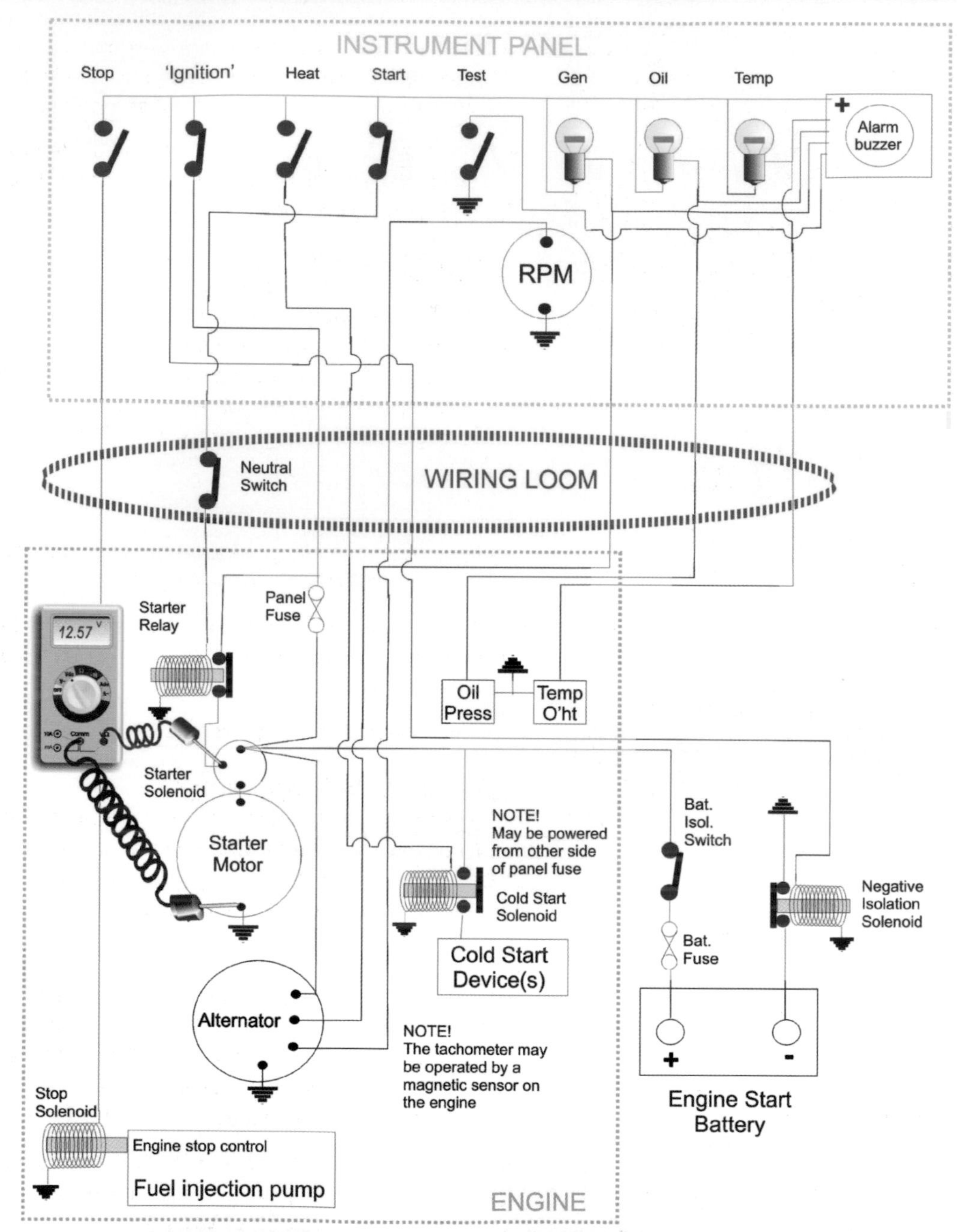

ENGINE ELECTRICS TROUBLESHOOTING
Starter switch 'ON'

Reproduced by permission of Volvo

- If the starter solenoid is not receiving current, you may be able to 'jump start' the engine. The instruments and warning lights may not then operate.
- If it is receiving current, you should be able to hear the solenoid click – if it does, then the contacts have failed or the engine/starter has seized.
- Older engines have a spiral groove along which the gear moves to engage the flywheel. This may need to be cleaned. If the starter just 'whirs', the starter pinion is sticking on its spiral thread. If it's stuck at the flywheel end and the starter won't turn, you can 'wind' it back to the starting position using a spanner on the front end of the starter shaft.

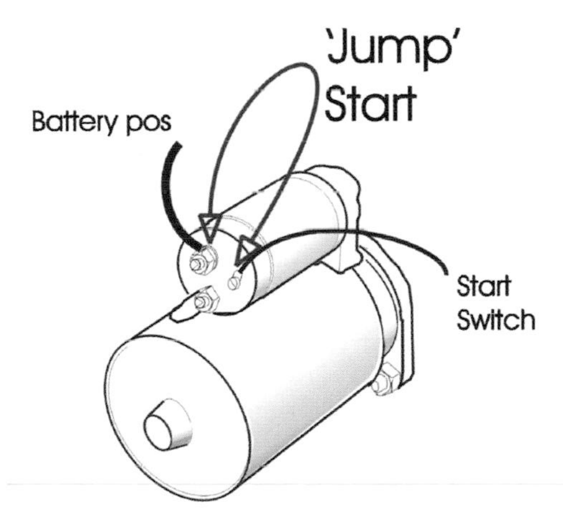

Jump starting

- Make sure that nothing you are wearing can get caught up in the machinery.
- Bridge the positive (battery) terminal on the starter solenoid and the starter switch terminal on the solenoid to turn the engine.
- This won't work if the solenoid has failed.

Engine turns but won't start

Turn off the cooling sea water cock in a sailing boat before continuing – turn on when engine starts. Don't run the starter motor for more than 30 seconds without an intervening five-minute cooling period.

Engine turns slowly

- Low battery state – parallel batteries if possible. However, don't connect a very discharged battery to a good one, as the low battery will pull the charged battery down and you may not be able to start. If possible, it's better to just connect to the 'good' battery.
- Check battery connections – remember the battery connection to the engine block and the starter motor bolts. These must all be clean and tight.
- If you have decompressers, decompress the cylinders while you turn the engine on the battery,

so that full turning speed is achieved, then close them (or only one if the system allows) to achieve a start.

Engine turns normally

- Check 'stop control' is not at 'stop'.

- If the engine is stopped electrically (no pull handle), the stop solenoid (which requires 12 volts to 'stop') may have jammed in the stop position. You should be able to move this to the 'run' position by hand.
- The stop mechanism may be mounted internally, in which case, remove the solenoid. You should then be able to operate the stop control in the engine.

- Some marinised engines (and generating sets) need 12 volts to run, so make sure the connections are sound and that 12 volts show at this point.

- Check fuel 'on'.
- Check fuel contents.
- Check 'cold starting procedure' is correct – read handbook!
- Check electric cold start system is functioning by following the troubleshooting wiring diagram.
- Bleed low-pressure fuel system – make sure it's fuel not water.
- Bleed high-pressure fuel system by loosening the high-pressure unions two and a half turns. Caution: high-pressure fuel can penetrate the skin.

- If you're out of fuel or there's a blockage, you could jury-rig a fuel supply.

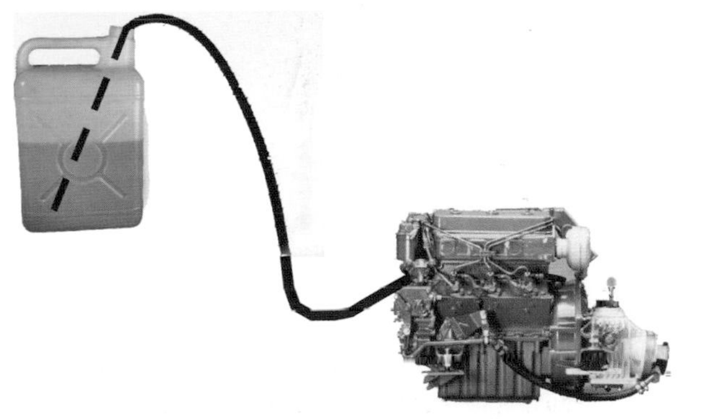

DO NOT USE STARTING FLUID ON A DIESEL – ENGINE DAMAGE IS LIKELY. No current diesel engine manufacturer approves the use of starting fluid. (If you must, spray some on a rag and hold the rag by the air intake as the engine is cranked. Get the engine overhauled!)

A much safer way is to apply heat to the intake area and pre-heat the air as it's sucked into the air intake. Remove the air cleaner and apply heat with a hairdryer. Using *exceptional care* to prevent fire and burning electrical circuits, a gas blow-lamp can also be used in an emergency, but only if you'll hit the rocks if you can't start the engine!

Remember that difficult starting from cold indicates a faulty technique, poor battery or connections or an engine problem, which can include a bent connecting rod caused by water in the cylinder.

NO RESPONSE TO THROTTLE

Generally, this is just a matter of a cable becoming disconnected.

- *Stop the engine* to ensure you don't get caught up in moving machinery.
- Check if the speed lever on the engine moves when the throttle is opened.
- If it does, the problem is at the fuel injection pump and an engineer will be required.
- It will be obvious at this stage if the cable is connected to the engine speed lever.
- If it is, then you'll need to look at the throttle control(s) and the dual-station box if fitted.

NO RESPONSE TO GEAR SELECTION

All drive types

It's possible that weed, net or some other fouling is reducing thrust to zero, though it's more likely that you would get at least some drive or visible disturbance of the water.

- *Stop the engine* to ensure you don't get caught up in moving machinery.
- Check that the clutch lever on the gearbox responds to movement of the gear selector and that *full* movement is achieved, as specified in the handbook/ workshop manual.
- Check the oil level in the gearbox – if clutch operation uses oil pressure, the clutch won't engage if there's no oil.
- If in doubt about the clutch lever operation over its full travel, disconnect the cable and operate the lever by hand and observe how far it moves to fully operate.
- Reconnect the cable and check that when the gear lever is moved, the clutch lever on the engine reaches full travel.

Shaft drive systems

If it does, start the engine.

- Ensure that there's nothing that you are wearing, long hair, etc., that can get caught up in moving machinery.
- Select forward idle (or astern if that was the problem).
- Check that the gearbox/prop shaft coupling is rotating.
- If it isn't, then the trouble is within the gearbox.
  – Volvo Penta 2000 series engines coupled to an MS2 gearbox have a known problem with the splined coupling on the gearbox input shaft. DB engineering have an excellent cure for this problem.
- If the coupling rotates, observe the prop shaft to see if this is rotating as well.
- If no rotation, then the problem is the attachment of the coupling to the prop shaft.
- Some couplings have pinch bolts, so check their tightness.

- Some couplings have, in addition, a bolt or roll pin through both the coupling and shaft, and this may have sheared. If this has happened, you'll need to replace the bolt or roll pin, but tightening the pinch bolts may get you home.
- If the prop shaft rotates, then the problem is at the propeller end.

Sail drive systems

All the rest of the system is invisible.

- The problem could be within the gearbox/leg.
- The problem could be that the propeller is not rotating or the propeller has fallen off.
- The boat will need to be lifted out of the water or someone will have to dive under the boat for further investigation.

Stern drive systems

By tilting the leg up sufficiently, you'll be able to see if the prop is still there or if it's rotating or not. However, some motor cruisers have such large bathing platforms that the props won't be visible.

NO WATER FROM EXHAUST

On many modern boats, the exhaust is not visible. If the exhaust is above the waterline, there will be a distinct change in exhaust note when the exhaust is dry. If the exhaust is below the waterline, then your first indication of a problem may be the engine overheat alarm sounding. In this case, the water pump impeller may already have been damaged. If you (or the engine or boat manufacturer) have fitted a water intake or exhaust overheat warning system, this will operate to tell you that the cooling water flow has failed.

- Stop the engine if safe to do so.
- After a short while, restart the engine. This allows time for any plastic sheeting that may have blocked the intake temporarily to float away.
- If the problem persists, check the weed strainer.

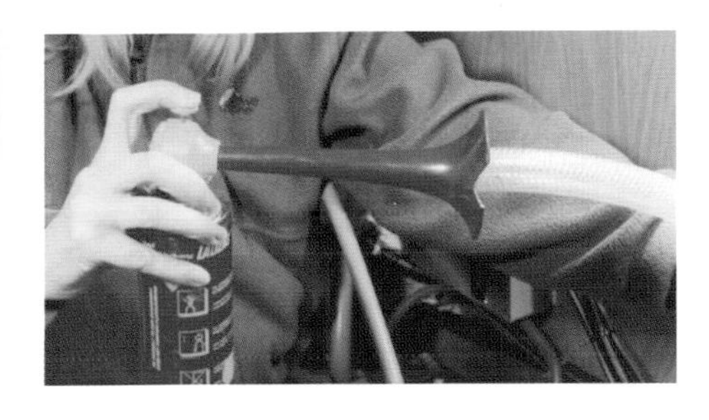

- Some weed strainers are fitted in such a way that you can clear an obstruction in the intake by 'rodding' with a suitable stick and the seacock open.
- Obstructions can also be removed using a gas fog horn or dinghy pump to blow down the intake pipe towards the open seacock.
- Check the raw water pump impeller and change if necessary.
- If the impeller looks intact, ensure that it's turning by blipping the starter quickly as you watch the impeller.
- At the same time, ensure that the water pump cam stays positioned correctly between the inlet and outlet pipes.

ENGINE OVERHEAT ALARM OPERATES

See also above if you can't observe the exhaust outlet.

- Stop the engine if it's safe to do so.
- Check the level of the water in the expansion tank, if fitted.
- Check for leaks in the cooling system.
- Wait for the system to cool sufficiently so that the water filler can be removed safely – use a rag to protect your hand as you do so and turn the cap very slowly to relieve any pressure safely. Steam will scald!
- Top up the cooling system if necessary using a mixture of freshwater and antifreeze. If the system is still hot, don't use cold water.
- Restart the engine and observe the temperature if you have a gauge. From cold, it may take ten minutes for the overheat alarm to be activated.
- Investigate why the water level was low.
- Modern wax-filled thermostats will fail open, so failure should not cause overheating.

- Older, bellows-type thermostats will fail closed, so overheating is likely.
- Don't run the engine with the thermostat removed, as engine damage due to undetected overheating is almost inevitable, despite what some books will tell you.

ENGINE LOW OIL PRESSURE ALARM OPERATES

- Stop the engine if safe to do so.
- Check the oil level.
- Top up if necessary.
- If the engine is running very hot, the oil pressure may fall due to the oil becoming too 'thin'.
- If the oil has been contaminated by diesel fuel – possibly a dribbling injector – the oil will become too 'thin'.
- A worn engine may have too low an oil pressure at idle when the engine is hot.
- The oil pressure relief valve may have stuck open.
- The oil pressure pump may be worn.
- The oil filter may have become clogged and doesn't have a bypass.

ALTERNATOR/GENERATOR WARNING LIGHT ILLUMINATES

Ideally, there will be a volt meter indicating system voltage and, better still, an ammeter as well.

- Check the alternator output by means of the volt meter and the ammeter if fitted.
- Advance the throttle from idle and check the light again.
- Note that some smart regulators have a 'soft' start, which allows the engine to warm up a little before the alternator starts to give any output.

- Stop the engine.
- Check the alternator (or dynamo/dynastart) drive belt.
- Check the connections on the alternator.
- If you have a smart regulator fitted, check the diagnostic lights if available.
- Call an engineer.

Maintenance

ENGINE SERVICING

The best guide to servicing is the engine handbook. Although service intervals are normally governed by the number of engine hours since the last service, for the leisure boater, it's more often dictated by the need to carry out a service at least annually.

Cooling system

- *Indirectly cooled* (freshwater cooling like a car) *engines* need to have the coolant replaced every second year. The coolant should be made up according to the engine handbook's instructions, but is often 50% water and 50% antifreeze. *Antifreeze* contains corrosion inhibitors, which are consumed over a period of time and must be replaced. Checking with an antifreeze tester will *not* test the inhibitors. Cylinder-head gasket failure is common on engines which do not have this done.

- *Raw water ('direct') cooled* (sea, lake or river water cooling) *engines* often have sacrificial zinc anodes to prevent electrolytic corrosion of internal parts. These must be checked annually and replaced if more than half has been consumed.

- The *heat exchangers* of 'direct' cooling systems may have anodes, which need to be checked. These will be found in the raw water section of the heat exchanger and may include oil and gearbox coolers and turbo inter-coolers as well. The handbook should tell you, but beware, if the handbook covers several different engine/gearbox configurations, only general advice may be given.

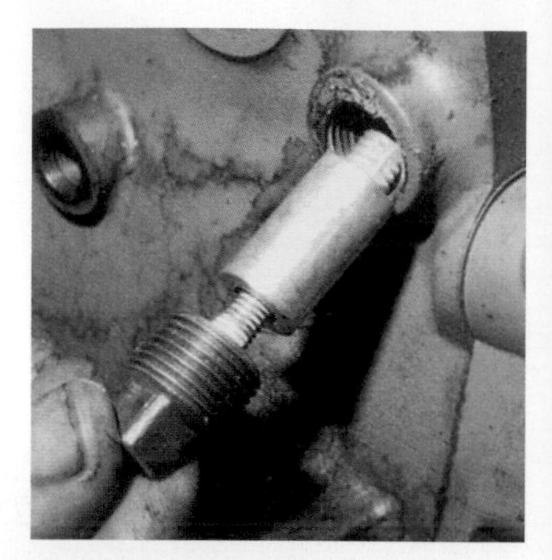

- It's generally recommended that the *raw water pump impeller* be changed annually.

- If the '*seacock*' is of the 'Blakes' type (Kingston Valve), this will need an annual service.
- If your engine is mounted close to the water line, it will probably have a *syphon break*. If it has a valve (it will have no pipe leading to a drain overboard), this must be serviced annually. Failure to do this can cause water to syphon into the internal working parts of the engine.

- Check the condition and security of all hose clips (jubilee clips).

Lubrication

- Change the oil on schedule, or more often if you have a lot of stop/start 'motoring'.
- Use the grade and viscosity of oil as indicated in the handbook – in the case of older engines, the

specified grade may not be obtainable, so use the nearest. In many sailing yachts, with minimal engine use, engines will often never become fully 'run in', even by the time the engine dies a natural death. In these situations, the use of 'synthetic' oil for engine lubrication, unless specified in the handbook, often compounds the problem.

Engine oil change

- Most people recommend that the engine should be run to warm up the oil, making its extraction easier.
- Let the engine stand for 10 minutes to allow the oil to settle.
- If the engine has a plumbed-in oil extraction pump, use this to extract the old oil.
- Otherwise, remove the dip-stick and insert the tube of the oil extraction pump, trying to get it as close to the bottom as possible.
- Pump out the oil.

- Pour the required quantity of new oil into the oil filler. (This won't be contaminated by the dirty oil

still contained in the old filter, but will give time for the oil to reach the sump).

- Using a suitable wrench, remove the old oil filter, trying to contain the spilled oil.
- Fit a new filter, first lubricating the oil seal, tightening as indicated on the instructions printed on the filter.

- Check the oil level. (As the filter doesn't yet contain any oil, it may over-read).

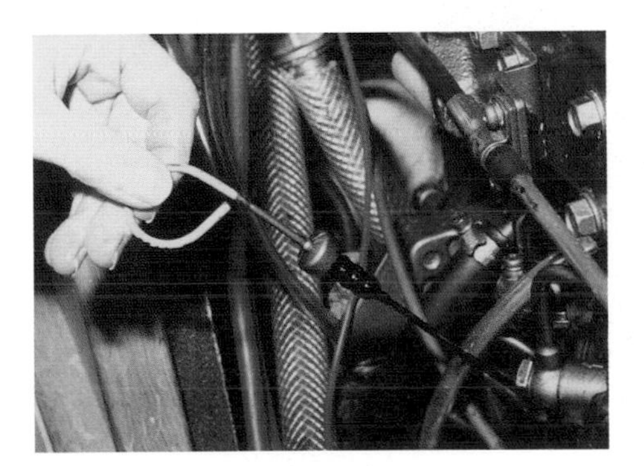

- Run the engine for a couple of minutes to check for leaks. If you have a mechanical 'stop' control, keep it pulled until the oil pressure light goes out, to allow oil pressure to build before the engine starts.
- Wait 10 minutes, check the oil level and top up if necessary.

Gearbox oil change

- Oil normally has to be removed by a pump, through the dip-stick hole.
- Refill with oil as specified – it may be the same specification as used in the engine or it may be completely different – sometimes the gearbox will have its own separate handbook.

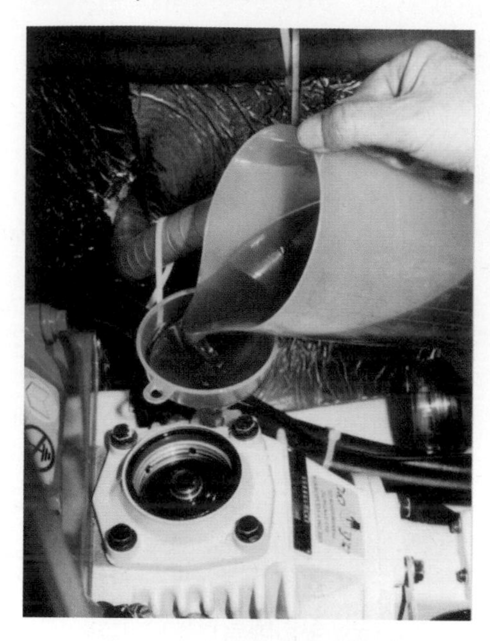

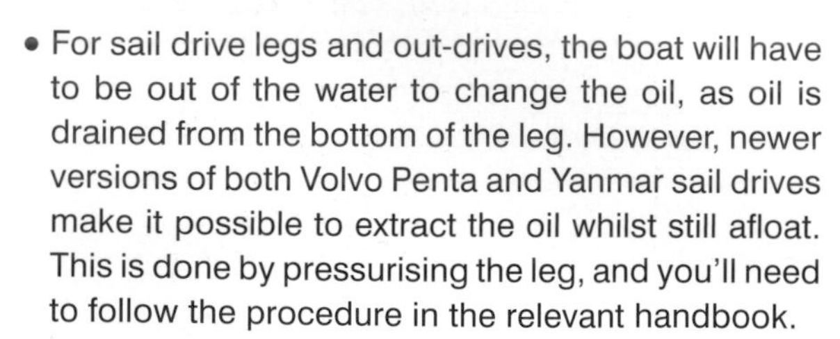

- For sail drive legs and out-drives, the boat will have to be out of the water to change the oil, as oil is drained from the bottom of the leg. However, newer versions of both Volvo Penta and Yanmar sail drives make it possible to extract the oil whilst still afloat. This is done by pressurising the leg, and you'll need to follow the procedure in the relevant handbook.

Fuel system

The *fuel tank* should be cleaned approximately every five years, but this is seldom done, because there's no drain plug and access is difficult. The dirt in the fuel filter bowl in the photograph indicates the need for tank cleaning.

- Turn off the *fuel cock*.

- Replace the filter element of the *pre-filter* and clean the water separating bowl.

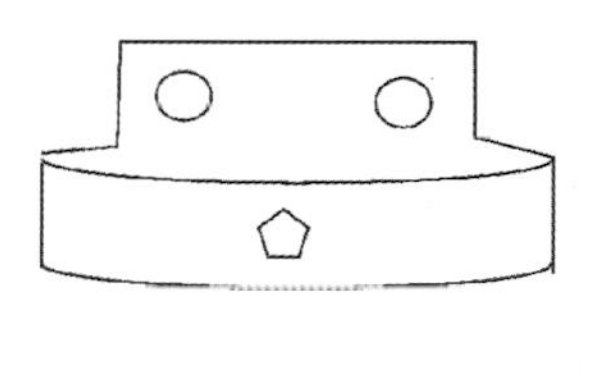

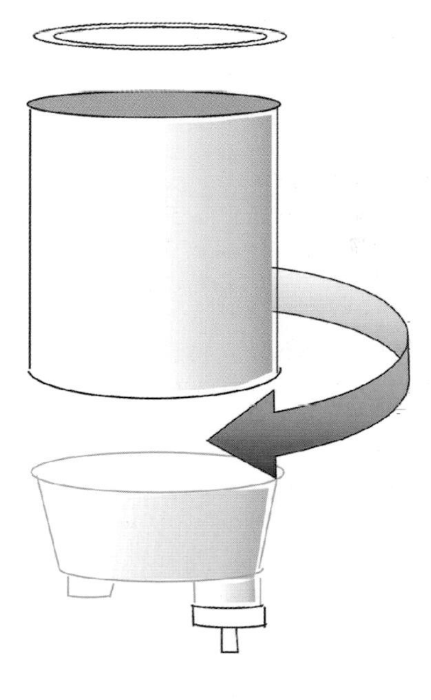

- Clean the filter element (if fitted) of the *fuel lift pump*.

- Replace the element of the engine *fine filter*.

- Open the fuel cock.

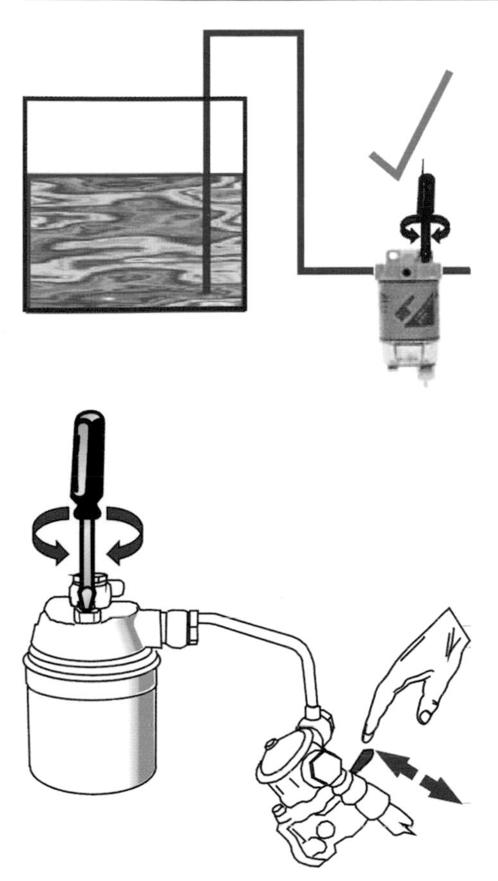

- If the fuel level in the tank is *above* the filter, bleed the filter by opening the bleed screw until fuel runs out, close the bleed screw.
- If the fuel level is *below* the filter, do not attempt to bleed.

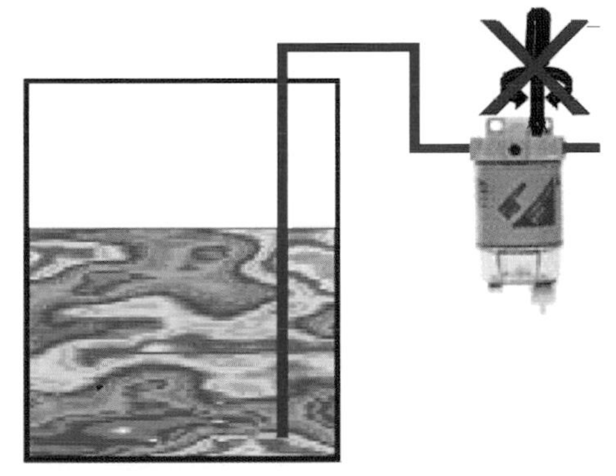

- *Bleed* the system.
- Service or replace the *fuel injectors* if symptoms dictate – light grey to blue smoke all the time and difficult starting. You may prefer to get an engineer to do this.

Air system

- Clean or replace the *air filters*, as indicated on the service schedule.

- Check the security of the air filter *housing*.

- If the air filter is connected to the engine with a *rubber hose*, check for softness and delamination.
- Work on the *turbo-charger*, if fitted, is best left to the experts.
- Check the condition of the exhaust injection bend.

Electrical system

- Check the cleanliness and security of all connections.
- Check the battery level and top up if necessary.

- Clean the battery case.
- Clean the battery terminals and apply petroleum jelly (Vaseline) to them.

ADJUSTING THE VALVE CLEARANCE

The engine maintenance schedule will require the valve gear to be adjusted at regular intervals. There is a small clearance between the valve operating gear and the valve stem. Too big a gap will lead to noise or even failure of the mechanism. Too small a gap can lead to the valve not closing properly, resulting in valve failure. Failure to check this can lead to serious engine damage. The engine handbook will give details of when this job needs to be undertaken.

Most marine engines have push-rod operated valves and these are simple to adjust. If your engine has twin overhead camshafts, valve clearance is more difficult to adjust and reference will have to be made to the engine's workshop manual.

- If you have a mechanical engine stop system, move the stop lever to 'stop'. Although this is not essential, as you will be turning the crankshaft by hand, there is an infinitesimal chance that the engine could start. So, if it's easy to do so, you may just as well prevent the engine starting, no matter how unlikely the event.

- As an alternative, or in addition, slacken off the fuel injectors by a couple of turns to prevent the engine from firing as it is turned by hand.
- On this engine (a Yanmar 3 GM30F), there's a breather that needs to be removed first, so that the rocker cover can be lifted clear. Remove the bolts.

- Move the breather out of the way.
- Undo the bolts holding the rocker box cover in place.

- Remove the rocker box cover exposing the valve gear.

- Identify the parts.

- Workshop manuals normally give instructions on how to position the crankshaft so that the valves are fully closed, so that they may be adjusted. Follow this if you have a manual. However, there is a simple method of ensuring that the valves of a particular cylinder are closed.

- Identify the inlet valve of one cylinder. This can be done by looking at the inlet and or exhaust manifolds and visually lining the duct with the valve.

Identifying the valves

- Put a ring spanner or a socket spanner onto the crankshaft nut so that you can rotate the engine.

- Check from the engine specification the direction of rotation and rotate the crankshaft until an inlet valve opens and closes. This indicates that that cylinder is just starting its compression stroke. Unless both valves are closed, air cannot be compressed, so this is a positive indication of valve closure. Rotate the crankshaft 180 degrees to the 'top of compression', which you can feel easily. Try it a few times so that you are confident you have found the correct point.

- Select the correct feeler gauge (the clearance is given in the engine manual, which also tells you if the procedure is carried out with the engine hot or cold).

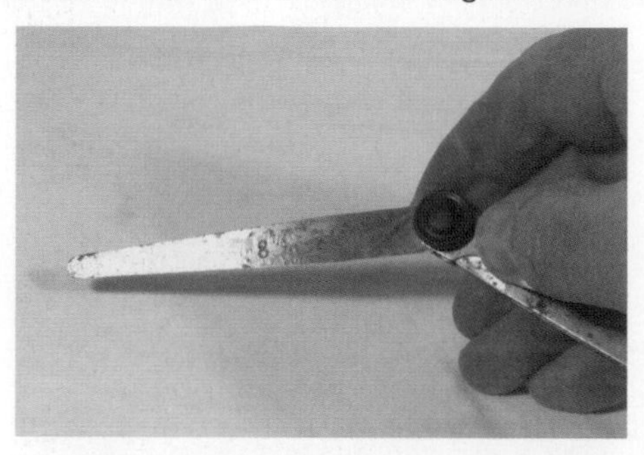

- Slide it into the gap between the adjuster and the push rod. You should feel a definite drag on the oiled gauge, but you should be able to get the gauge in the gap without difficulty.
- If the gap needs to be adjusted, slacken the lock nut, screw the adjuster in or out to get the correct gap and retighten the lock nut. When you retighten the lock nut, grip the screwdriver firmly to prevent the adjuster turning with the nut.

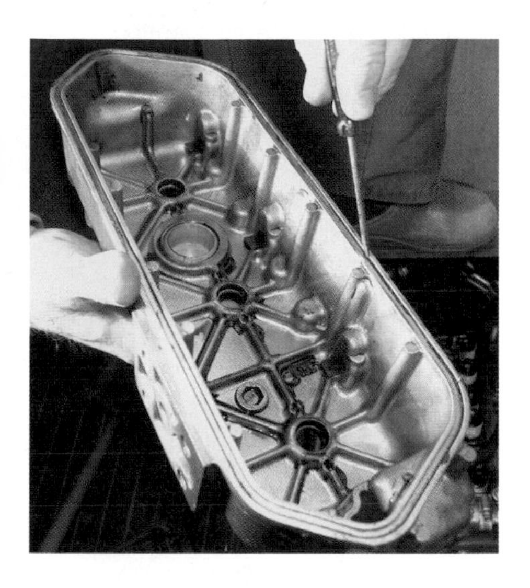

- Recheck the clearance and readjust if necessary.
- Clean the mating surfaces of the rocker box cover. Ensure that no dirt gets into the valve gear and the top of the cylinder head.
- Check the condition of the sealing gasket and replace if necessary.
- Apply a suitable non-setting gasket sealant.

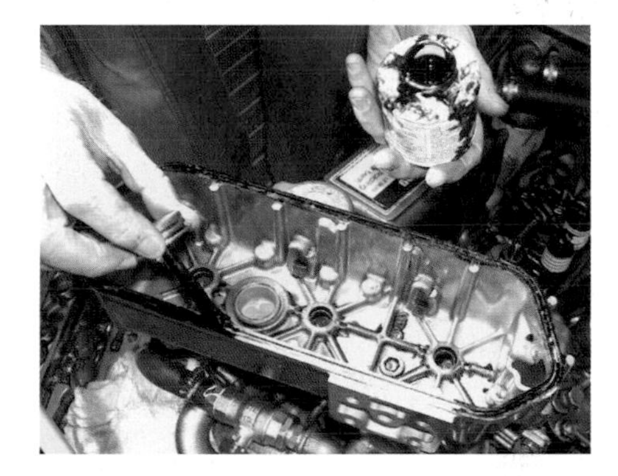

- Replace the rocker box cover and any other components that have been removed.

Tip

Volvo Penta 2000 series engines have a crankcase breather hole in the rocker box cover gasket. This must be aligned with the breather hole in the cylinder head.

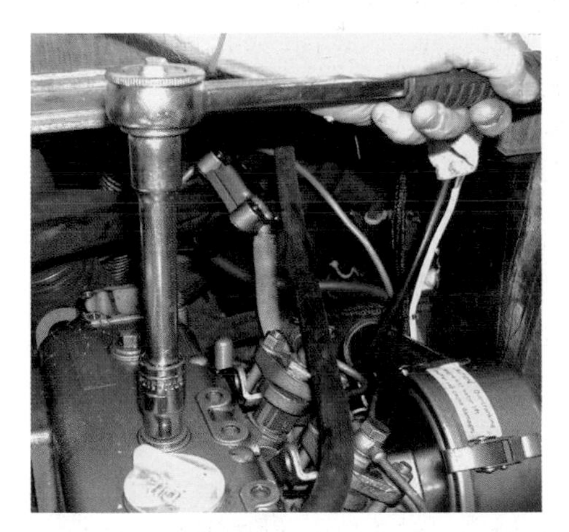

ENGINE MOUNTS

Modern engines are mounted on flexible feet to reduce noise and vibration. These mountings can deteriorate or fail completely, so regular checking of their condition is required. The rubber mounts will deteriorate rapidly if contaminated by oil or diesel fuel, and delamination of the rubber from the metal can occur.

The mounting normally comprises a metal bracket bonded onto a rubber flexible block, which, again, is bonded onto a metal bracket. The mounting is attached to the engine's foot and to the boat's structure. Normally there are four mounts on an engine, two each side.

Engine mounts need to be checked for settling, which will affect engine alignment and vibration.

Engine mounting brackets need to be checked for fractures.

- Fractures may occur due to ageing.

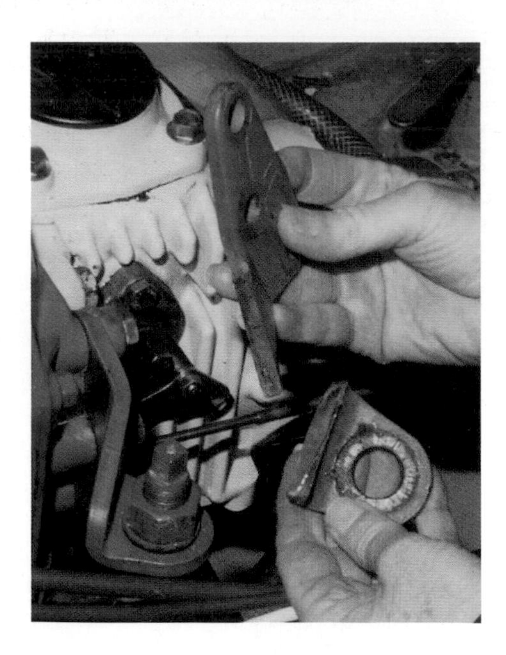

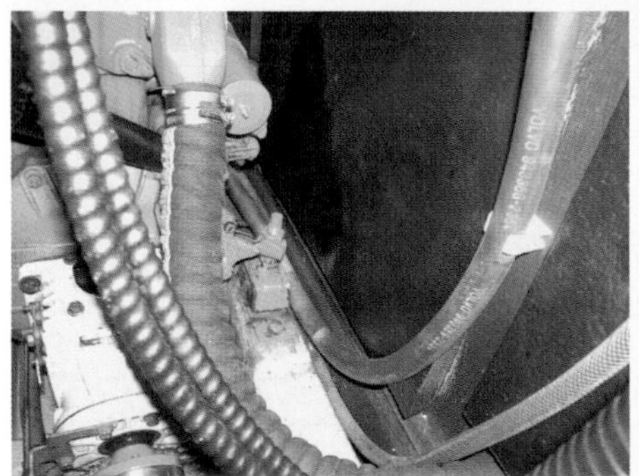

- Rust around a welded stud indicates full or partial failure. The weld in the photograph was completely fractured and the engine could be lifted from it.
- Fractures may also occur due to fouling the propeller with a rope.
- The rubber may become unbonded from the metal.

Sail drive installations

Sail drive units normally have only three mounts. There's a pair at the 'front' of the engine (where the drive belt(s) is/are) and one at the 'back' of the gearbox. These mountings sit on a fibreglass plinth bonded to the hull. In some installations, the engine sits 'back to front', with the leg reversed on the bottom of the gearbox. Here, you'll find the gearbox at the 'front'. The mount at the gearbox end is different from the other pair.

Deteriorating mounts will cause vibration but won't affect engine alignment. Fractured mounts can cause the engine to move and then the integrity of the diaphragm seal is at risk.

Shaft drive installations

With four mounts, the front pair will often have a different elasticity from the rear pair, as the weights sitting on them will differ. They may also be different in design.

The prop shaft is normally rigidly mounted, though there will be some elasticity at the front end, where the stern gland is mounted in a rubber hose or moulding. Most builders will fit a rigid coupling between the gearbox and the prop shaft. The engine mounts allow movement of the engine, so the stern gland must be able to withstand this movement.

Deteriorating engine mounts will cause vibration, and sag will cause the engine to become misaligned with the prop shaft. Broken mounts will allow the engine to bounce and it may then jump off its bearer. This will cause a heavy load on the stern gland, which may fail, or, in the case of some modern seals, displacement of the face seal can cause a large leak, putting the boat at risk of sinking.

Alignment

If a rigid coupling is fitted, alignment of the engine and the prop shaft is critical. Misalignment will cause vibration and wear in the gearbox bearings and the cutless bearing.

Semi-flexible couplings will allow up to two degrees angular misalignment, and fully flexible couplings allow up to eight degrees. The actual misalignment allowed depends on rpm and engine torque. Vertical and horizontal alignment needs to be accurate in all cases.

Winterisation

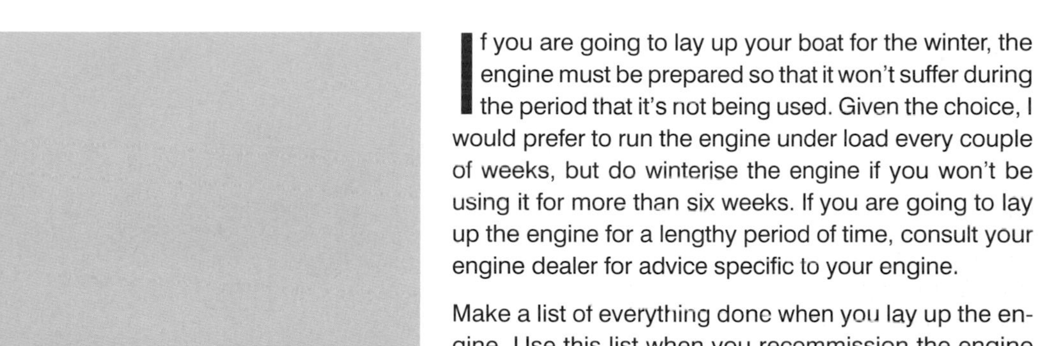

If you are going to lay up your boat for the winter, the engine must be prepared so that it won't suffer during the period that it's not being used. Given the choice, I would prefer to run the engine under load every couple of weeks, but do winterise the engine if you won't be using it for more than six weeks. If you are going to lay up the engine for a lengthy period of time, consult your engine dealer for advice specific to your engine.

Make a list of everything done when you lay up the engine. Use this list when you recommission the engine to ensure everything is reinstated correctly.

FUEL TANK

Keep the tank full, especially during the winter, to minimise condensation in the tank.

SERVICING

Carry out a full service of the engine and gearbox as required by their handbooks.

COOLING SYSTEM

Note, you will be running the exposed engine. Ensure that neither you nor your clothes become entangled in rotating machinery.

- If you have direct cooling, run the engine until the thermostat is open. If you don't have a temperature gauge, feel the thermostat housing – when it's hot, the thermostat is open.

- Stop the engine, close the seacock and disconnect the hose, so that you can dip it into a bucket of freshwater. Restart the engine and when the water reduces to about 25%, add a litre of antifreeze. As the bucket empties, stop the engine. Do not restart the engine until it's recommissioned. This process replaces the seawater with an engine-friendly liquid.
- Don't remove the water pump impeller. Change the impeller when you recommission, or you may lose some of the water/antifreeze mix.

- If you have an indirect cooling system, it's not a bad idea to flush the seawater part as above, but the thermostat doesn't have to be open – because it's in the freshwater part of the system. Seawater remaining in the system could freeze in very low temperatures, and draining the system isn't always fully effective. If there's a gearbox and engine oil cooler, these will have seawater in them and will now be protected by the antifreeze.

AIR SYSTEM

- Disconnect the exhaust hose from the engine and check the condition.
- Remove the air cleaner(s) from the engine.
- Spray water repellent oil up the exhaust and air intakes and then seal with plastic bags to prevent damp air entering.
- Check the exhaust injection bend for a build up of carbon and for any cracking due to corrosion.
- Check the exhaust pipe hose for a build up of carbon and for delamination.
- If the boat is to remain afloat, block the exhaust where it leaves the hull, or reconnect the exhaust to prevent water flooding the boat, otherwise leave it open to ventilate the hull.

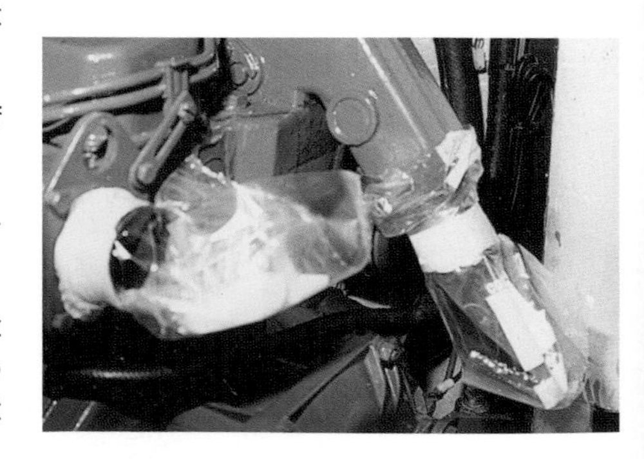

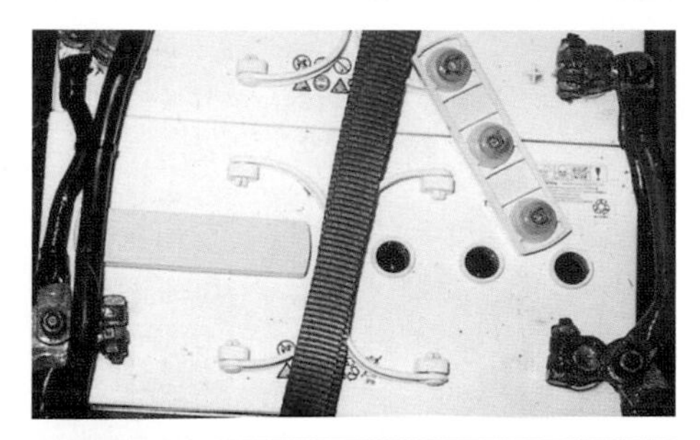

ELECTRICS

- Check the battery level.
- Top up the batteries with distilled water if necessary.
- The battery needs to be recharged monthly, so take it home if necessary.
- Clean and tighten all electrical connections – if in doubt, renew them – remember the negative cables on the engine block.

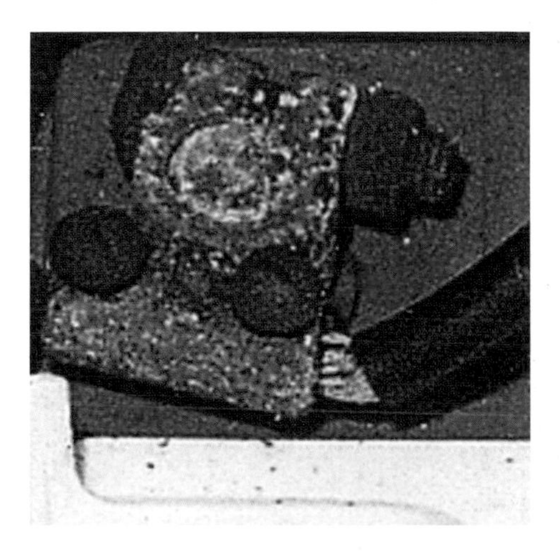

CLEANLINESS

- Clean the engine and engine compartment.
- Remove any rust and then touch up paint.

- Spray the engine with a water displacing fluid such as Boeshield T.9 – include the wiring and terminals. There have been reports that WD 40 has caused damage to the insulation of low-grade, non-marine wiring, although I have not encountered the problem myself.

VENTILATION AND HEATING

To help prevent condensation, open all access panels to the engine compartment to allow ventilation.

If mains power is available:

- use a low-powered tubular heater in the engine compartment;
- if possible, run a dehumidifier – although relatively expensive to buy, weekly running costs are not great and you will be protecting your investment.

IF THE BOAT IS TO BE LAID UP OUT OF THE WATER

Depending on the type of drive unit:

- Check and renew all anodes as necessary. This may entail removing the propeller(s).
- Clean the propeller(s).
- Antifoul propellers if that is your choice action.
- Prime and antifoul the stern gear. If the stern gear is aluminium, ensure that any antifouling paint used is compatible.
- Ensure that raw water cooling intakes are unobstructed.

Recommissioning

Ideally you would have made a list of everything that you did when you laid up the engine, noting those items that needed action when recommissioning. Work methodically through your list, which will vary depending on the make of engine, gearbox and drive system.

- It's essential to verify the watertight integrity of the vessel prior to relaunch.
- Some stern glands, such as the Volvo Penta, need to be squeezed to remove any trapped air immediately after launching.
- Test run the engine and check all systems prior to proceeding 'to sea'.
- Engine alignment should be checked about two weeks after launching to allow the hull to settle to its natural shape afloat.

HORIZONTAL ALIGNMENT

Unless the engine has been moved by disconnecting the mounting bases from the engine bearers, horizontal alignment (along the axis of the boat in plan view) will not change. Even then, if the same design of mounts and the same attachment holes are used, alignment will not be compromised seriously.

In the event that horizontal alignment has to be checked and adjusted, it's a matter of trial, error and patience, with the engine being levered sideways a tiny bit at a time, back and front.

CHECKING ANGULAR AND VERTICAL ALIGNMENT

1. Select forward or astern, as appropriate, to lock the prop shaft.
2. Slacken off the drive coupling bolts by about one turn.

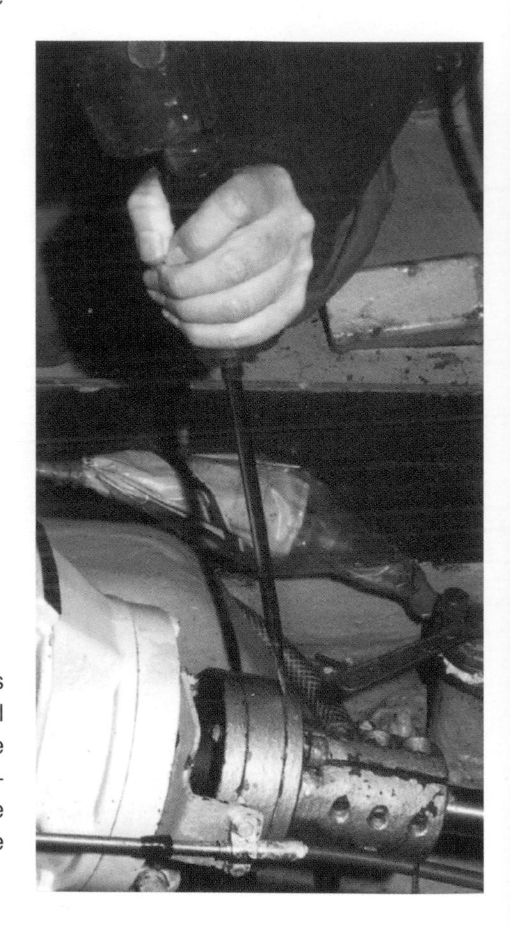

3. Select neutral.
4. Tap the shaft coupling to make sure it frees from the gearbox output and takes up its natural position. You may need to slacken the bolts some more and use a paint scraper or something similar to prise the coupling apart. Next, retighten the bolts until the coupling flanges touch at just one point.
5. Make a table like that shown.
6. Rotate the shaft so that one bolt is at the top and mark the coupling so that you know the starting point. Measure the clearance between the gearbox

ENGINE ALIGNMENT

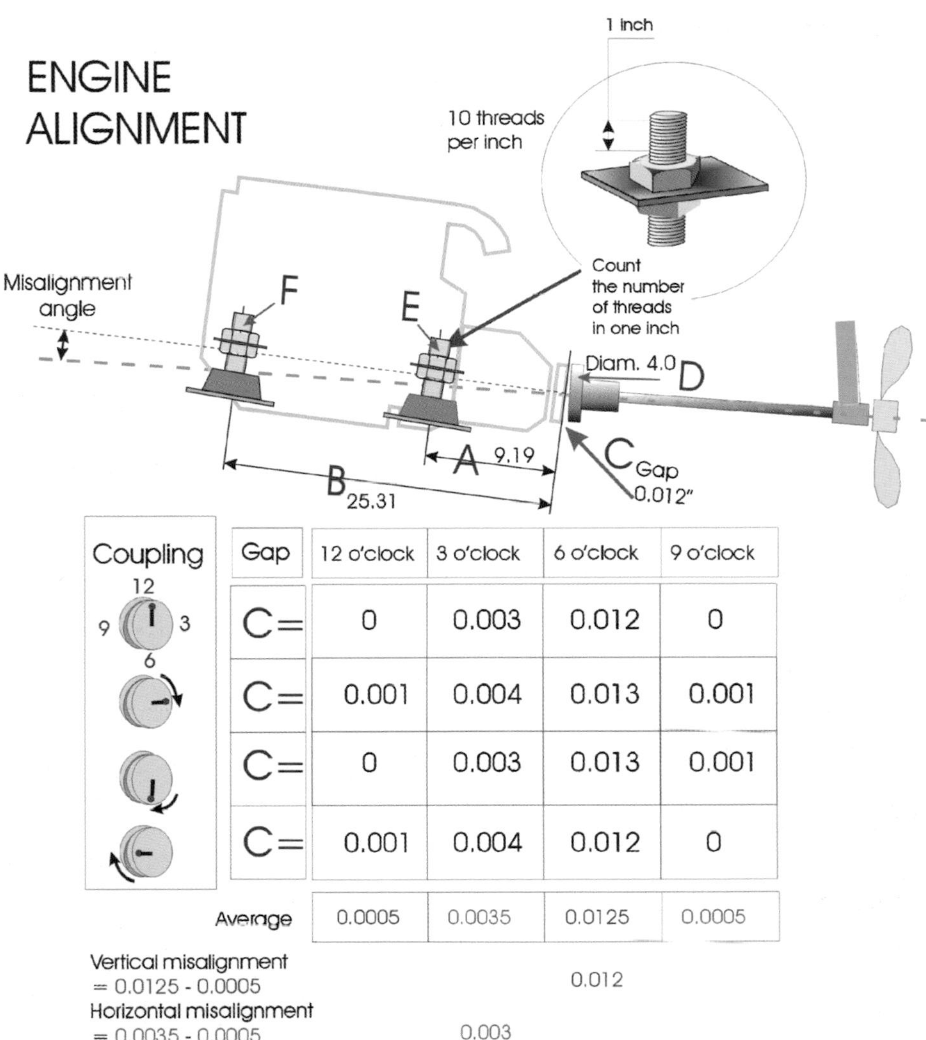

Coupling	Gap	12 o'clock	3 o'clock	6 o'clock	9 o'clock
12 / 9 / 3 / 6	C=	0	0.003	0.012	0
	C=	0.001	0.004	0.013	0.001
	C=	0	0.003	0.013	0.001
	C=	0.001	0.004	0.012	0
	Average	0.0005	0.0035	0.0125	0.0005

Vertical misalignment
= 0.0125 - 0.0005   0.012

Horizontal misalignment
= 0.0035 - 0.0005   0.003

Diameter of coupling	4.0	
Feeler gauge size of gap	0.012	("12 thou.")
Coupling to rear engine mount	9.19	
Coupling to front engine mount	25.31	
Number of threads per inch	10	

$$\text{Adjustment E} = \frac{C \times A}{D} = \frac{0.012 \times 9.19}{4.0} = 0.0275$$

$$\text{Adjustment E} = \frac{C \times B}{D} = \frac{0.012 \times 25.31}{4.0} = 0.0759$$

Adjustment E = Threads per inch x 0.0275 = 0.20 TURNS LOWER
Adjustment F = Threads per inch x 0.0759 = 0.76 TURNS LOWER

drive plate and the coupling plate using a feeler gauge at the 12, 3, 6 and 9 o'clock positions. Enter in the table.

7. Rotate the shaft 90 degrees, measure and note the clearances again.
8. Do the same for 180 and 270 degrees, noting the clearances.
9. Add the four figures and divide by four for each 'clock position', using the table.
10. For a four-inch (100-mm) diameter coupling, the resulting average should be no greater than 0.004 inches (0.1 mm).
11. If all is well, retighten the coupling bolts.

Adjusting the alignment

1. Measure the distances as shown in the photograph.

2. Enter the distances and coupling clearances into the table.
3. Measure the number of threads per inch (cm) of the engine mount adjustment bolts to obtain the thread pitch, as shown in the alignment diagram.
4. Enter the thread pitch into the table.
5. Calculate the number of turns for both the forward and aft mounts using the alignment table.
6. Slacken the mounting lock nuts.

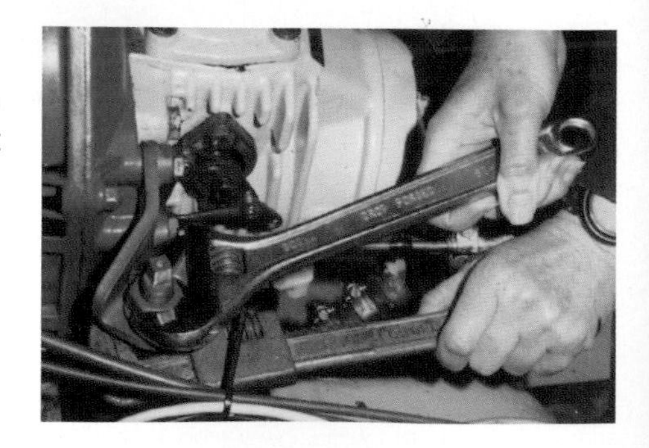

7. Raise or lower the lower nuts the calculated number of turns.
8. Tighten the lock nuts.

9. Recheck the clearance.

Engine Compartment Fire Extinguishers

THE CAUSES OF FIRE IN ENGINE COMPARTMENTS

On diesel engine boats used for leisure purposes, the most likely cause of engine compartment fires is an electrical short circuit or electrical component overheating. It is therefore essential that any electrical system running through the engine compartment can be isolated from outside. It is rare for the fuel system itself to cause the fire, and normally the fuel system will become involved only if a fuel pipe burns through. However, if a fuel leak does occur, leaking fuel can come into contact with a hot component. A broken high-pressure fuel injection pipe allowing fuel to spray onto a hot turbo-charger casing or uncooled exhaust can cause a fierce fire in a very short space of time.

FIRE DETECTION

The thing that worries me most about engine installations in recreational sailing and motor yachts is the fact that there's normally no fire detection system. Ask yourself the question: 'How would I know that I have an engine fire?' The answers I get from my students are that you would see the smoke, the floor would get hot, there would be smoke in the exhaust. None of these are necessarily true. The real answer is that you wouldn't know until the fire was well established.

A fully fitted system will have a fire warning (sounder and light) and either manual or automatic operation of the fire extinguisher, and may incorporate an engine shut-down system.

Owners could fit a very simple and cheap fire detector in the form of a temperature switch and a sounder. If the compartment gets too hot, the buzzer will alert you. Don't be tempted to fit a domestic smoke alarm in your engine bay. You will almost certainly be plagued by false alarms.

TYPES OF EXTINGUISHER

Not all systems are appropriate for smaller yachts and motor cruisers because of weight, space and cost considerations, but the following can be used.

Dry powder

Dry powder is effective against fires of flammable liquids, including spray fires. Powders:

- Are capable of effecting very rapid extinguishment, but provide little cooling effect and are ineffective once the powder has settled.
- Settle out after use and present the problem of post-fire clean up. Where it isn't possible to fully clean up the residue, such as in an engine compartment, there are concerns about the long-term corrosive effects.
- May not go a round corners into less accessible spaces.
- Can cause internal damage when entering a running engine, needing at best the cylinder head to be removed.

- Are not recommended for use in occupied spaces because they are unpleasant to breathe and obscure visibility.

Halocarbon gas systems

A number of fire extinguishing halocarbon gases with zero ozone depletion potential (ODP) have been developed. These include both HFCs (hydrofluorocarbons) and PFCs (perfluorocarbons). These new agents share many of the characteristics of the old Halons, but PFCs have been banned for marine applications by the IMO (International Maritime Organisation).

Although HFCs have zero ozone depletion potential, they have some global warming potential. They can be used in manned spaces, unlike Halon. They are full-flood systems that fill the whole protected space with the fire suppression medium.

Fine solid particulate technology

Aerosol and inert gases are formed pyrotechnically. The solid aerosol acts directly on the flame, cooling it; the gases deliver the aerosol to the fire. Solid particulates have very high effectiveness to weight ratios. They are not suited for use in manned areas because the particulate material would be inhaled and the visibility severely reduced. Careful use and installation is required because of the extremely high burning temperature.

These may be automatic, electrically or manually operated.

DIFFERENT TYPES OF SYSTEM

Hand-held extinguisher discharged into the engine space from outside

If you haven't an automatic system, this is your only way of tackling an engine compartment fire. Only an extinguisher suitable for use on electrical and liquid fuel should be used, and this will generally be of the dry powder type for recreational craft use. Access must be via a very small hole, so that extra oxygen is not admitted to the engine space, and many boats don't have one. Opening the engine compartment

will cause the fire to flare, as more oxygen is admitted, so if this is the method you use, you need to make a small access hole and close it with a small plug that is quick to remove. This also begs the question of how you know you have a fire in the first place.

Whatever the regulations say, I don't consider this to be a satisfactory way of fighting an engine compartment fire, although the ability to use one, through a small access hole, if the dedicated extinguisher fails is a good idea.

Extinguisher mounted inside the engine space and discharged automatically

This is a self-contained and common system. On smaller boats, the extinguisher has a small glass vial that melts or fractures in the presence of high temperature, allowing the escape of the extinguishant under pressure. This is the minimum and cheapest system you should aim for. There are also extinguishers with fire sensing wire, a combustible fuse or a tube that burns through at the point of fire, connected to and containing the extinguishant.

This still begs the question of how you know you have a fire. The engine may 'cough' as the extinguisher operates and starves the engine of air, but it may then continue to run. An automatic engine shut-down system can be installed, but see below.

Extinguisher system mounted inside the engine space and discharged either automatically or manually from outside

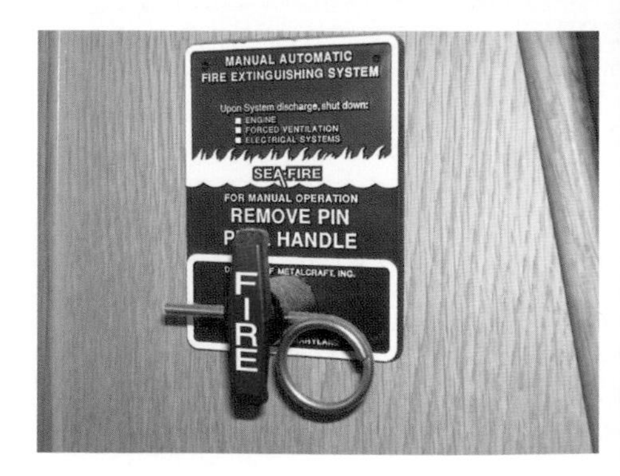

This system will warn of the fire and operate the extinguisher(s). You have the option of manual discharge if necessary. Some systems will also shut down the engine(s) and the associated electrical systems, but have an override allowing engine restart if necessary.

These are the most versatile and expensive systems.

AUTOMATIC OPERATION

How automatic should the system be? There are two distinct schools of thought about this.

In big ships, a detection system alerts the crew, the crew decides what to do, the space is evacuated and the extinguishing system is operated. If the fire is localised, the fire is attacked at that point manually by the crew.

In aircraft, the system is never automatic, the crew is alerted and then takes the appropriate action according to the circumstances.

In small boats, there is sometimes an automatic fire extinguisher in the engine compartment. Some of the newly introduced systems follow this automatic operation with automatic disconnection of the engine electrical system, but others give a fire warning but with electrical activation by the crew and no automatic shut down.

In my opinion, the first requirement of the system is to alert the crew. This should be standard on all boats. This warning device could operate the extinguisher automatically (or automatic extinguisher operation could alert the crew), or allow the crew to operate the system manually. Whether the electrical system should also be shut off is open to debate. On most marine diesels with electric stop solenoids, you need electrical power to stop the engine. If the power is removed, you may not be able to stop the engine if you wish. The corollary to this is that most diesel engine room fires are of electrical origin, so one of the first actions should be to switch off the power. A recent fire on board a new motor cruiser was tackled by the automatic fire extinguisher. The powerful twin engines continued to run and the extinguishant was sucked straight through the engines and the fire was not put out. Automatic engine shut down would have ensured that the fire was extinguished.

You should also turn off the fuel supply from the tank, so this must be accessible from outside the engine compartment.

It's unlikely the builder of a sailing boat will have installed any system at all. Whatever system you opt for, ensure that you follow the manufacturer's instructions, both on the installation itself and the size of the extinguisher compared to the volume of the compartment.

MY SYSTEM

I decided to protect my valuable investment by fitting a halocarbon gas system with automatic and manual operation, a warning system but no automatic engine shut down. My insurance company requires that there's an automatic extinguisher in my fuel tank space. On my boat this is the deep cockpit locker, which also contains my Eberspacher heater, so the requirement is entirely sensible.

This satisfied my requirements that:

- I should know that I had an engine compartment fire;
- If I took no action, the extinguisher would operate;
- I could operate the system manually should I need to;
- I could stop the engine myself.

This was in addition to a simple fire warning system consisting of a temperature switch and buzzer, plus a small fire extinguisher hole that I had fitted previously.

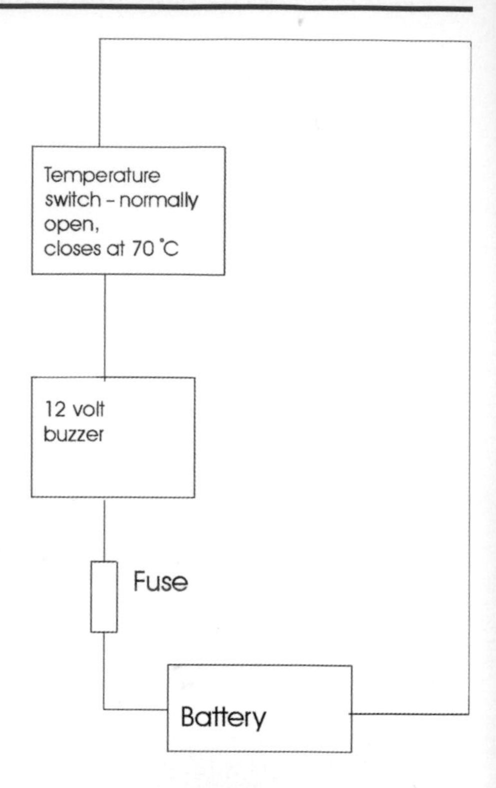

DIY Fire Warning Circuit

Tools

> **Note!**
>
> If your engine is older, you will need a set of A/F or even Whitworth and BSF spanners.

There should always be a basic toolkit on the boat. This should include:

- tools for routine engine servicing;
- tools for routine boat work.

Some of these tools will be common to both categories. Additionally, a more extensive kit will be needed according to the amount of work you intend doing yourself.

ENGINE TOOLS

- Set of open-ended spanners (6 mm–17 mm).
- Set of ring spanners (10 mm–17 mm).
- Set of socket spanners (6 mm–17 mm).
- Adjustable spanners.
- Self-gripping wrench.
- Set of flat-blade screwdrivers.
- Set of cross-head screwdrivers.
- Pliers.
- Long-nose pliers.

- Side-cutting pliers.
- Set of Allen keys (hexagonal wrenches).
- Set of feeler gauges.
- Ball pein hammer.
- Hide or rubber hammer.
- Hobby knife.
- Oil filter wrench.
- Sump pump.
- Magnet on a stick.
- Mirror on a stick.
- Torch (preferably a head torch).

ENGINE SPARES AND CONSUMABLES

- Self-amalgamating tape.
- Insulating tape.
- Cable ties.
- Liquid gasket.
- Thread-locking compound.
- PTFE thread-sealing tape.
- Water-displacing fluid.
- A range of jubilee clips (hose clamps).
- Spare coolant hose.
- Spare fuel hose.
- Spare fuses.
- Grease.
- Vaseline (petroleum jelly).
- Engine service spares.
- Primary fuel filter.

- Engine fuel filter.
- Oil filter.
- Drive belts.
- Engine oil.
- Gearbox oil.
- Raw water pump impeller and gasket.

Index

Picture Credits

The following images have been reproduced from other sources with permission.

Early diesel engine on page 2. Reproduced by permission of Volvo.

Cutaway Yanmar engine on page 2. Reproduced by permission of E. P. Barrus Ltd.

Petter Mini 6 engine on page 2. Reproduced by permission of Petter.

Diaphragms image on page 8. Reproduced by permission of E. P. Barrus Ltd.

Hot exhaust pipes on page 18. Reproduced by permission of Volvo.

Detroit diesel engine on page 19. Reproduced by permission of Detroit Inc.

Photo of honing on page 40. Reproduced by permission of Volvo.

Power curves diagrams on pages 44 to 48. Reproduced by permission of E. P. Barrus Ltd.

Exploded fuel pump on page 60. Reproduced by permission of Perkins Engines Company Limited.

Fuel lift pump on page 62. Reproduced by permission of Perkins Engines Company Limited.

Injection pump on page 65. Reproduced by permission of Perkins Engines Company Limited.

Fuel injection pump on page 66. Reproduced by permission of Volvo.

Pipe clamps on page 66. Reproduced by permission of Volvo.

Injection pipes on page 66. Reproduced by permission of Volvo.

Air-cooled diesel engine on page 78. Reproduced by permission of Petter.

Tube stack on page 93. Reproduced by permission of E. J. Bowman (Birmingham) Ltd.

Heat exchanger unit photo on page 93. Reproduced by permission of E. J. Bowman (Birmingham) Ltd.

Cooling water header tank photo on page 93. Reproduced by permission of E. J. Bowman (Birmingham) Ltd.

Exploded heat exchanger photo on page 95. Reproduced by permission of E. J. Bowman (Birmingham) Ltd.

Exploded anti-syphon valve on page 98. Reproduced by permission of Volvo.

Thermostat testing on page 102. Reproduced by permission on Volvo.

Exploded starter diagram on page 154. Reproduced by permission of E. P. Barrus Ltd.

Nightrider on page 268. Reproduced by permission of David Gibby.

You rely on us.
Can we rely on you?